Managing Microsoft Project Online

Rolly Perreaux, PMP, MCSE, MCT

Version 17.12.04

EXCLUSIVELY PUBLISHED BY

PMO Logistics
679 Roberta Avenue
Winnipeg, Manitoba, Canada R2K 0K9

Copyright © 2017 by Roland Perreaux

All rights reserved. No part of the contents of this document may be reproduced or transmitted in any form or by any means without written permission of the publisher.

Information in this document, including URL and other Internet Web site references, is subject to change without notice. Unless otherwise noted, the example companies, organizations, products, domain names, e-mail addresses, logos, people, places, and events depicted herein are fictitious, and no association with any real company, organization, product, domain name, e-mail address, logo, person, place, or event is intended or should be inferred. Complying with all applicable copyright laws is the responsibility of the user. Without limiting the rights under copyright, no part of this document may be reproduced, stored in or introduced into a retrieval system, or transmitted in any form or by any means (electronic, mechanical, photocopying, recording, or otherwise), or for any purpose, without the express written permission of PMO Logistics Inc.

The names of manufacturers, products, or URLs are provided for informational purposes only and PMO Logistics makes no representations and warranties, either expressed, implied, or statutory, regarding these manufacturers or the use of the products with any Microsoft technologies. The inclusion of a manufacturer or product does not imply endorsement of Microsoft of the manufacturer or product. Links are provided to third party sites. Such sites are not under the control of PMO Logistics and PMO Logistics is not responsible for the contents of any linked site or any link contained in a linked site, or any changes or updates to such sites. PMO Logistics is not responsible for webcasting or any other form of transmission received from any linked site. PMO Logistics is providing these links to you only as a convenience, and the inclusion of any link does not imply endorsement of PMO Logistics of the site or the products contained therein.

PMO Logistics may have patents, patent applications, trademarks, copyrights, or other intellectual property rights covering subject matter in this document. Except as expressly provided in any written license agreement from PMO Logistics, the furnishing of this document does not give you any license to these patents, trademarks, copyrights, or other intellectual property.

PMO Logistics, Professional Training Series, Upgrading Skills Series, TriMagna Corporation and TriMagna Corporation logo are either registered trademarks or trademarks of PMO Logistics Inc. in Canada, the United States and/or other countries.

Microsoft, Active Directory, Internet Explorer, Outlook, Project Server, SharePoint, SQL Server, Visual Studio, Windows and Windows Server are either registered trademarks or trademarks of Microsoft Corporation in the United States and/or other countries.

All other trademarks are property of their respective owners.

Author: Rolly Perreaux, PMP, MCSE, MCT

Publisher: PMO Logistics
Developmental Editor: Heather Perreaux
Cover Graphic Design: Andrea Ardiles
Technical Testing: Underground Studioworks

Post-Publication:
Errata List Contributors:

Introduction i

Table of Contents

Introduction

Table of Contents i
About the Author iii
Introduction iv
Facilities v
About This Course vi
Prerequisite Skills vii
Course Outline viii
Course Requirements x
Document Conventions xi

Module 1: Deploying Microsoft Project Online

Module Overview 1-1
Lesson 1: Installing Microsoft Project Online 1-2
   Practice: Creating a New Project Online Subscription 1-11
Lesson 2: Working with Office 365 Admin Center 1-13
   Practice: Working with Office 365 Admin Center 1-11

Module 2: Managing Security

Module Overview 2-1
Lesson 1: Overview of Project Server Security 2-2
Lesson 2: SharePoint Security Permissions 2-7
   Practice: Working with SharePoint Security Permissions 2-12
Lesson 3: Project Online Security Permissions 2-14
   Practice: Creating Security Templates 2-22
Lesson 4: Creating Project Online Security Entities 2-25
   Practice: Working with Project Online Security Principals 2-35

Module 3: Deploying Project Clients

Module Overview 3-1
Lesson 1: Overview of Project Clients 3-2
   Practice: Installing Project Online Professional 3-7
Lesson 2: Configuring Project Clients 3-9
   Practice: Configuring Project Online Professional 3-16
Lesson 3: Using Project Web App 3-18
   Practice: Using Project Web App 3-25

Module 4: Configuring Project Online

Module Overview 4-1
Lesson 1: Configuring Time and Task Management Settings 4-2
   Practice: Configuring Time and Task Management Settings 4-15
Lesson 2: Configuring Operational Policies 4-19
   Practice: Configuring Operational Policies 4-26
Lesson 3: Importing Resources and Project Plans 4-27
   Practice: Importing Resources and Projects 4-34

Module 5: Configuring Enterprise Data Settings

Module Overview 5-1
Lesson 1: Configuring Enterprise Custom Fields 5-2
    Practice: Configuring Enterprise Custom Fields 5-11
Lesson 2: Configuring Enterprise Objects 5-17
    Practice: Configuring Enterprise Objects 5-22

Module 6: Customizing Project Sites

Module Overview 6-1
Lesson 1: Working with Project Sites and Elements 6-2
    Practice: Working with Project Sites 6-8
Lesson 2: Creating a Custom Project Site Template 6-9
    Practice: Creating a Custom Project Site Template 6-15

Module 7: Project Online Administration

Module Overview 7-1
Lesson 1: Working with Project Online Workflows 7-2
    Practice: Customizing Workflows and PDPs 7-13
Lesson 2: Sharing Project Online with External Users 7-22
    Practice: Sharing Project Online with External Users 7-30
Lesson 3: Managing Queue Jobs and Enterprise Objects 7-33
    Practice: Working with Enterprise Objects 7-37
Lesson 4: Troubleshooting Tools 7-39

Hands-On Lab: How to Create a Project Online Power BI Center

Create the Power BI Center using SharePoint Online 1
Sign Up for a Power BI Account 2
Using the Power BI Project Online Content Pack 6
Upgrading Free Power BI account to Power BI Pro 8
Adding Power BI Reports to a SharePoint Page 8
Modifying the Power BI Center Home Page 11
Sharing the Power BI Center Site 13
Sharing the Power BI Dashboard and Testing 14

About the Author

Rolly Perreaux, PMP, MCSE, MCT

Rolly Perreaux is a Business Solutions Architect and Founder for PMO Logistics, a company that specializes in Project Portfolio Management consulting services and training.

Rolly has over 30 years of hands-on work experience with a unique and complementary blend of business, technology, project management and finance experience. He has successfully developed and implemented many Business Solutions and IT Management programs. He has a Business Administration diploma and designations from the PMI, Microsoft, Compaq, IBM, CheckPoint and CompTIA.

Rolly specializes in the following Microsoft technologies: Project Online/Server, SharePoint, Office 365, Dynamic 365, Azure, VSTS, Teams, Planner, Power BI, PowerApps and Flow.

Introduction

- **Name**
- **Company affiliation**
- **Title/function**
- **Job responsibility**
- **Your experience with:**
  - Project Online/Server
  - SharePoint Online/Server
  - Office 365
- **Your expectations for the course**

The instructor will ask you to introduce yourself by providing the information on the slide to the other students in the class.

Facilities

- Class hours
- Building hours
- Parking
- Restrooms
- Meals
- Phones
- Messages
- Smoking
- Recycling

You will be given information pertaining to class hours, building hours, parking, restrooms, meals phones messages, smoking, and recycling.

- Be aware of the location of exits for your own safety in the event of an emergency. Your instructor may give you specific instructions if necessary.

- Please turn off cell phones and pagers.

- Please refrain from using e-mail during class time; check e-mail only during breaks.

- To get the most out of this class, your active participation is encouraged.

About This Course

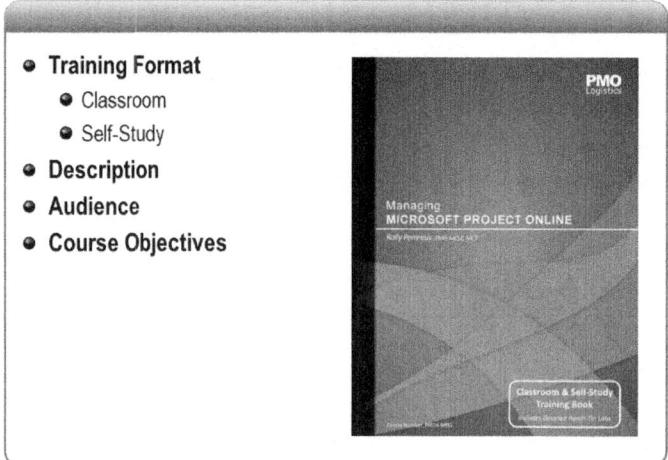

Training Format

This course is designed to work in either an instructor-led classroom or as a self-study.

Description

The goal of this two-day course is to provide students with the knowledge and skills necessary to effectively plan, deploy and manage Microsoft Project Online.

Audience

This course is intended for Administrators, Systems Engineers, PMO Managers, Project Managers, Consultants and other people responsible for the deployment and management of a Microsoft Project and Portfolio Management (PPM) Solution using Project Online.

Course Objectives

After completing this course, you will be able to:

- Deploy Project Online.
- Work with Office 365 Admin Center
- Configure and manage security.
- Install and configure Project clients.
- Configure and manage time and task management settings.
- Create enterprise custom fields and lookup tables.
- Configure and manage time and task management settings.
- Customize project sites.
- Import projects and resources.
- Create and configure Project Online workflows.
- Share Project Online with external users.
- Work with troubleshooting tools.
- How to create a Project Online Power BI Center.

Prerequisite Skills

> - **Working knowledge of the following:**
>   - Internet Web Browser
>   - Microsoft Project Professional
>   - Basic project management concepts

This course requires that you should have a working knowledge of the following:

- Internet Web Browser
- Microsoft Project Professional.
- Basic project management concepts.

Course Outline

- Module 1: Deploying Microsoft Project Online
- Module 2: Managing Security
- Module 3: Working with Microsoft Project Clients
- Module 4: Configuring Project Online
- Module 5: Configuring Enterprise Data Settings
- Module 6: Customizing Project Sites
- Module 7: Project Online Administration

Bonus Hands-On Lab
- How to Create a Project Online Power BI Center

Module 1: Deploying Microsoft Project Online

This module describes the features and how to install Microsoft Project Online in Office 365. Lessons presented in this module will detail how to create a new Project Online subscription and how to add a service to an existing subscription. You will also learn how to work in the Office 365 Admin Center by adding new users, assigning product licenses and creating and assigning groups to users.

Module 2: Managing Project Server Security

This module describes how to manage Project Server security. Lessons presented in this module will cover SharePoint and Project Server permission modes and detail how security works in both permission modes. You will also learn how to create Project Server security users, groups and categories and configuring the various permissions.

Module 3: Working with Microsoft Project Clients

This module describes how to install on-premises and cloud-based project clients. Lessons presented in this module will detail how to install both Project clients and create Project Online profiles. You will also learn how to access and sign in to Project Web App and how to impersonate users as part of testing and troubleshooting Project Web App.

Module 4: Configuring Project Online

This module describes how to configure and manage time and task management settings. Lessons presented in this module will detail how time and task reporting work in Project Server, what are financial periods and timesheet settings and how to configure them. You will also learn how to configure Operational Policies and import resources and project plans to Project Online.

Module 5: Configuring Enterprise Data Settings

This module describes how to configure the enterprise data settings. Lessons presented in this module will cover how to create and configure enterprise custom fields, and lookup tables. You will also learn about the Resource Breakdown Structure and configuring enterprise calendars and the Enterprise Global Template.

Module 6: Customizing Project Sites

This module describes how to configure, manage and customize Project Sites. Lessons presented in this module will explain how the elements of a Project Sites and how they work together. You will also learn how to create a custom Project Site template to be used in Enterprise Project Types (EPT).

Module 7: Project Online Administration

This module describes the administrative tasks that a Project Online Administrator will be required to know. Lessons presented in this module will detail how to configure Project Online Workflows, including Phases, Stages, PDPs and EPTs. You will learn how to share Project Online with external users and how to manage Queue Jobs and enterprise objects. You will also learn the various tools available to troubleshoot your Project Online deployment.

Bonus: How to Create a Project Online Power BI Center

This article provides the step-by-step instructions on how to create and share a custom Project Online Power BI Center in your organization's production environment.

Course Requirements

- **Desktop Computer or Laptop with:**
  - Microsoft Windows 7, 8.1 or 10 operating system
  - High Speed Internet connection to Microsoft Office 365 and tenant sites
- **Download the course file at:**
  - https://tinyurl.com/PRS16-MPO
- **To complete Module 7-Practice: Sharing Project Online with External Users, you will require one of the following:**
  - Microsoft Account (Skype, Outlook.com, OneDrive, Windows 10, Windows Phone, and Xbox LIVE)
  - Office 365 Account
  - Azure AD Account

This section provides the equipment requirements to successfully complete the course.

Equipment

In this course, you will need a desktop computer or laptop with the following requirements to perform the hands-on practices:

- Microsoft Windows 7, 8.1 or 10 operating system
- High Speed Internet connection to Microsoft Office 365 and tenant sites

Course Files

Please download the course file at: https://tinyurl.com/PRS16-MPO

Notice

To complete Module 7-Practice: Sharing Project Online with External Users, you will require one of the following:

- Microsoft Account (Skype, Outlook.com, OneDrive, Windows 10, Windows Phone, and Xbox LIVE)
- Office 365 Account
- Azure AD Account

NOTE: You can create a disposable Outlook.com account for purpose of completing the practice.

Document Conventions

Convention	Use
Bold	Represents commands, command options, and syntax that must be typed exactly as shown. It also indicates commands on menus and buttons, dialog box titles and options, and icon and menu names.
Italic	In syntax statements or descriptive text, indicates argument names or placeholders for variable information. Italic is also used for introducing new terms, for book titles, and for emphasis in the text.
Title Capitals	Indicate domain names, user names, computer names, directory names, and folder and file names, except when specifically referring to case-sensitive names.
ALL CAPITALS	Indicate the names of keys, key sequences, and key combinations —for example, ALT+SPACEBAR.
`monospace`	Represents code samples or examples of screen text.
[ ]	In syntax statements, enclose optional items. For example, [filename] in command syntax indicates that you can choose to type a file name with the command. Type only the information within the brackets, not the brackets themselves.

The following conventions are used in course materials to distinguish elements of the text.

Convention	Use
Bold	Represents commands, command options, and syntax that must be typed exactly as shown. It also indicates commands on menus and buttons, dialog box titles and options, and icon and menu names.
Italic	In syntax statements or descriptive text, indicates argument names or placeholders for variable information. Italic is also used for introducing new terms, for book titles, and for emphasis in the text.
Title Capitals	Indicate domain names, user names, computer names, directory names, and folder and file names, except when specifically referring to case-sensitive names. Unless otherwise indicated, you can use lowercase letters when you type a directory name or file name in a dialog box or at a command prompt.
ALL CAPITALS	Indicate the names of keys, key sequences, and key combinations —for example, ALT+SPACEBAR.
`monospace`	Represents code samples or examples of screen text.
[ ]	In syntax statements, enclose optional items. For example, [filename] in command syntax indicates that you can choose to type a file name with the command. Type only the information within the brackets, not the brackets themselves.

This page is intentionally left blank

Module 1: Working with Microsoft Project Online

Contents

Module Overview 1
Lesson 1: Installing Microsoft Project Online 2
    Features of Microsoft Project Online 3
    Creating a New Project Online Subscription 4
    Adding a Service to an Existing Subscription 9
    Practice: Creating a New Project Online Subscription 11
Lesson 2: Working with Office 365 Admin Center 13
    Adding New Users to Office 365 14
    Assigning Product Licenses to Users 16
    Creating Groups in Office 365 17
    Assigning Office 365 Groups to Users 18
    Practice: Working with Office 365 Admin Center 19
Summary 23

EXCLUSIVELY PUBLISHED BY

PMO Logistics
679 Roberta Avenue
Winnipeg, Manitoba, Canada R2K 0K9

Copyright © 2017 by Roland Perreaux

All rights reserved. No part of the contents of this document may be reproduced or transmitted in any form or by any means without written permission of the publisher.

Information in this document, including URL and other Internet Web site references, is subject to change without notice. Unless otherwise noted, the example companies, organizations, products, domain names, e-mail addresses, logos, people, places, and events depicted herein are fictitious, and no association with any real company, organization, product, domain name, e-mail address, logo, person, place, or event is intended or should be inferred. Complying with all applicable copyright laws is the responsibility of the user. Without limiting the rights under copyright, no part of this document may be reproduced, stored in or introduced into a retrieval system, or transmitted in any form or by any means (electronic, mechanical, photocopying, recording, or otherwise), or for any purpose, without the express written permission of PMO Logistics Inc.

The names of manufacturers, products, or URLs are provided for informational purposes only and PMO Logistics makes no representations and warranties, either expressed, implied, or statutory, regarding these manufacturers or the use of the products with any Microsoft technologies. The inclusion of a manufacturer or product does not imply endorsement of Microsoft of the manufacturer or product. Links are provided to third party sites. Such sites are not under the control of PMO Logistics and PMO Logistics is not responsible for the contents of any linked site or any link contained in a linked site, or any changes or updates to such sites. PMO Logistics is not responsible for webcasting or any other form of transmission received from any linked site. PMO Logistics is providing these links to you only as a convenience, and the inclusion of any link does not imply endorsement of PMO Logistics of the site or the products contained therein.

PMO Logistics may have patents, patent applications, trademarks, copyrights, or other intellectual property rights covering subject matter in this document. Except as expressly provided in any written license agreement from PMO Logistics, the furnishing of this document does not give you any license to these patents, trademarks, copyrights, or other intellectual property.

PMO Logistics, Professional Training Series, Upgrading Skills Series, TriMagna Corporation and TriMagna Corporation logo are either registered trademarks or trademarks of PMO Logistics Inc. in Canada, the United States and/or other countries.

Microsoft, Active Directory, Internet Explorer, Outlook, Project Server, SharePoint, SQL Server, Visual Studio, Windows and Windows Server are either registered trademarks or trademarks of Microsoft Corporation in the United States and/or other countries.

All other trademarks are property of their respective owners.

Author: Rolly Perreaux, PMP, MCSE, MCT

Publisher: PMO Logistics
Developmental Editor: Heather Perreaux
Cover Graphic Design: Andrea Ardiles
Technical Testing: Underground Studioworks

Post-Publication:
Errata List Contributors:

Module Overview

- Installing Microsoft Project Online
- Working with Office 365 Admin Center

In this module, you will learn the features and install Microsoft Project Online. You will also learn about the Office 365 Admin Center and how to manage Office 365 users and groups as they are the foundation for a successful portfolio management (PPM) solution with Project Online.

Objectives

After completing this module, you will be able to:

- Install Microsoft Project Online
- Work with Office 365 Admin Center

Lesson 1: Installing Microsoft Project Online

- Features of Microsoft Project Online
- Creating a New Project Online Subscription
- Adding a Service to an Existing Subscription

Microsoft Project Online brings together the business collaboration platform services of SharePoint Online with structured execution capabilities to provide flexible work management solutions. Project Online unifies project and portfolio management (PPM) to help organizations align resources and investments with business priorities, gain control across all types of work, and visualize performance through powerful dashboards.

This lesson provides an overview of the features of Project Online, how to create new subscription and how to add a service to a subscription.

Objectives

After completing this lesson, you will be able to:

- Explain the features of Microsoft Project Online
- Create a New Project Online Subscription
- Add a Service to an Existing Subscription

Features of Microsoft Project Online

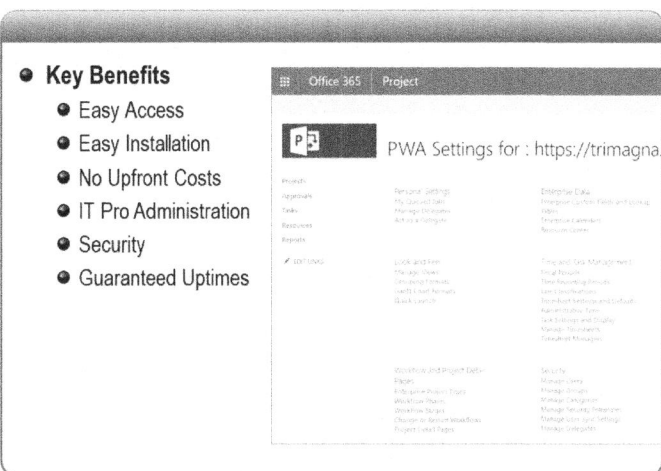

Microsoft Project Online is a flexible online solution for PPM and everyday work. Delivered through Office 365, Project Online enables organizations to get started, prioritize project portfolio investments and deliver the intended business value from virtually anywhere on nearly any device.

As an integral part of Office 365, Project Online provides <u>almost all</u> project and portfolio management functions via your browser. This new cloud-based solution offers many of the same features and benefits as locally installed Project Server 2016 on-premises.

Project Online provides the following key benefits:

- **Easy Access** – Executives, portfolio managers, project managers, project resources, team members, and stakeholders are able to access the Project Online system from anywhere provided an internet connection can be accessed.

- **Easy Installation** – Deployment of Project Online happens in minutes.

- **No Upfront Infrastructure Costs** – Project Online has no required upfront infrastructure costs. Customers do not have to purchase servers, or worry about Windows Server, SQL Server, or other server licenses.

- **IT Pro Administration** – All operational maintenance is handled through the Microsoft Office 365 Datacenter. There is no longer a requirement to commit IT resources to tasks such as updates, disaster recovery, and maintenance. Also, preventive maintenance scripts are run on your databases to prevent problems before they occur.

- **Security** – Office 365 employs robust disaster recovery capability, globally-redundant back-ups, and extensive privacy features. Filters help protect users against spam and viruses.

- **Guaranteed Uptimes** – Microsoft offers guaranteed, financially-backed uptimes and phone support 24 hours a day, seven days a week.

Creating a New Project Online Subscription

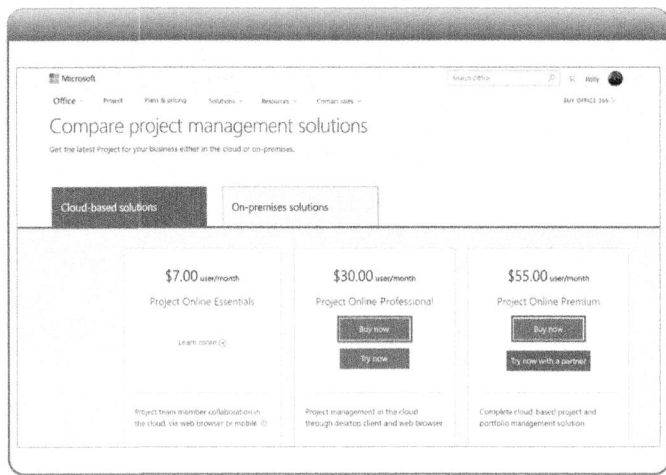

Creating a new Microsoft Project Online subscription is a very simple process.

1. Launch your web browser and type https://www.microsoft.com/project
2. From the **Project** web page, click **See products & pricing**.

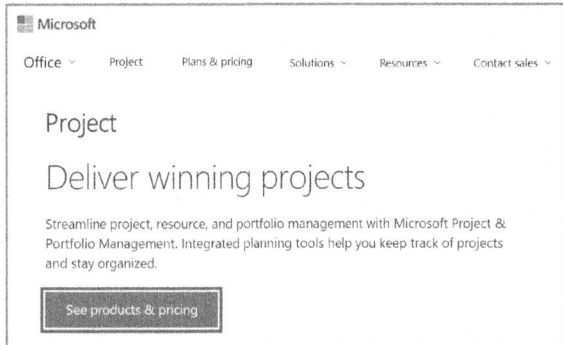

3. On the **Compare project management solutions** page, click **Buy now** for either Project Online Professional OR Project Online Premium.

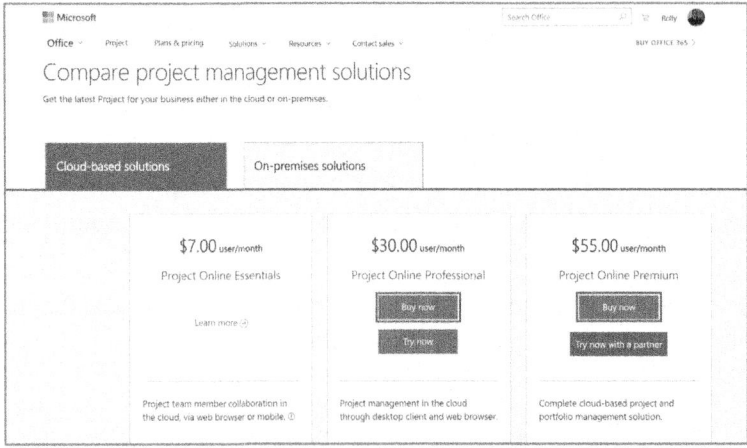

4. On the **Welcome, let's get to know you** page, complete all fields and click **Next**.

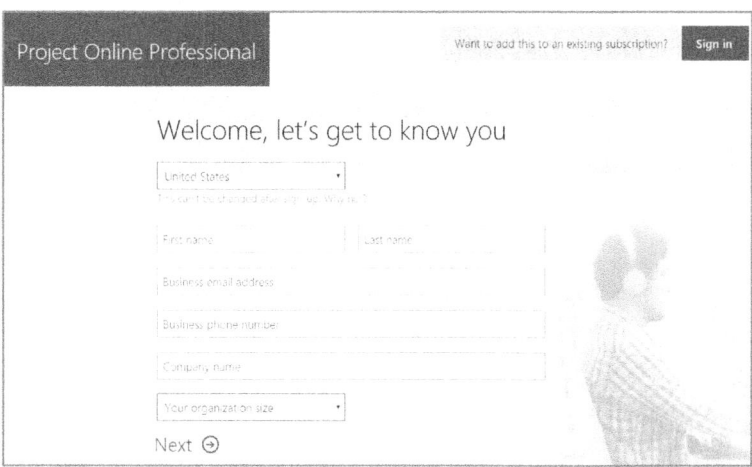

5. On the **Create your user ID** page, complete the form and click **Create my account**.

NOTE: Select the .onmicrosoft.com domain if you already own your domain name.

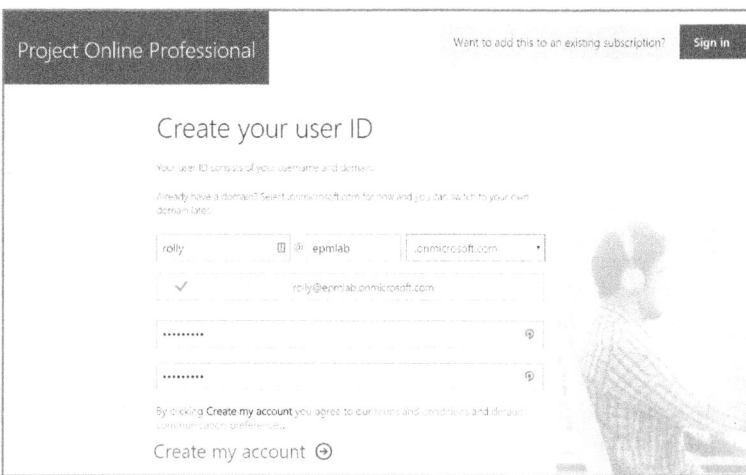

6. On the **Save this info. You'll need it later** page, copy the information and click **You're ready to go...**

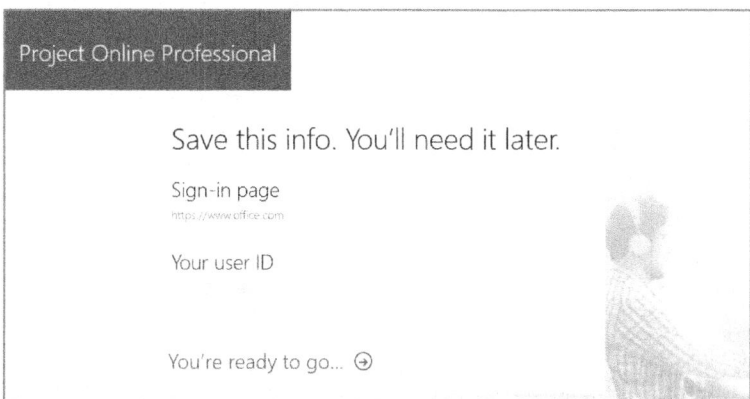

7. On the **Office 365** page, under the **Apps** section, click **Admin**.

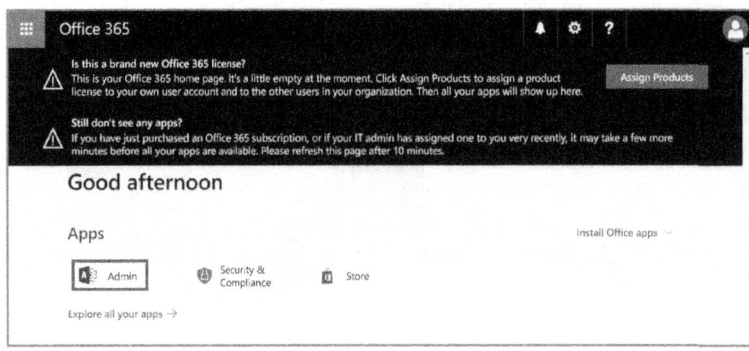

8. On the **Admin center** page, in the **Welcome to the admin center** window, click **Next** and continue through a series of windows and then click **Finish**.

9. On the **Admin center** page, click **Go to setup**.

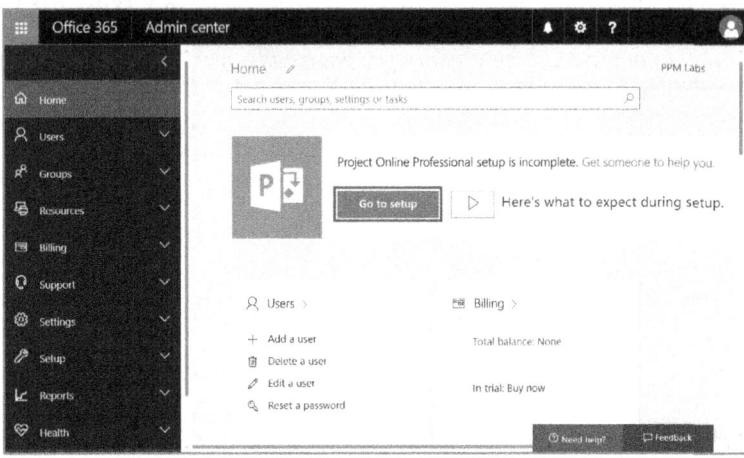

10. On the **Add new users** page, add your users that will be using Project Online Professional and click **Next**.

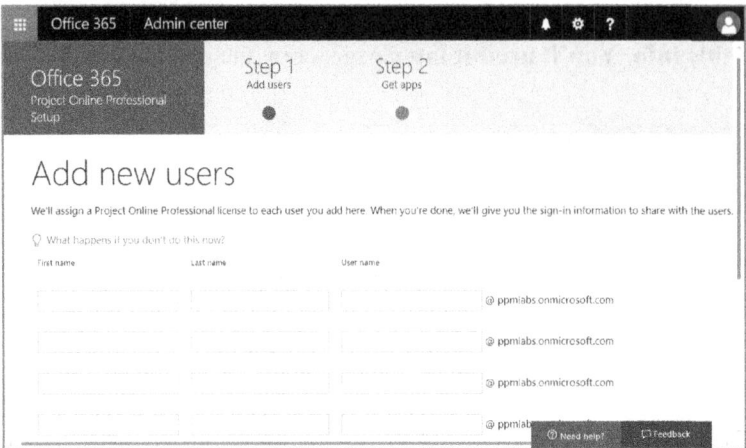

11. On the **Install your Office apps** page, if you want to install **Project Pro for Office 365**, click **Install now** OR if you want to install the software later, click **Next**.

Module 1: Working with Microsoft Project Online   1-7

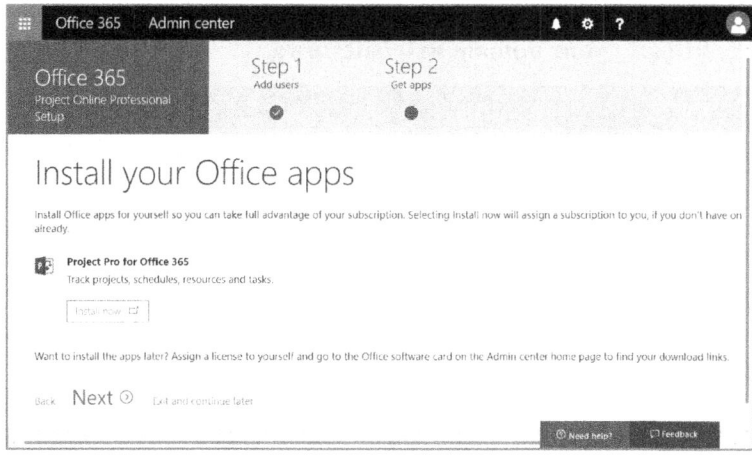

12. On the **You've reached the end of setup** page, click **Go to the Admin center**.

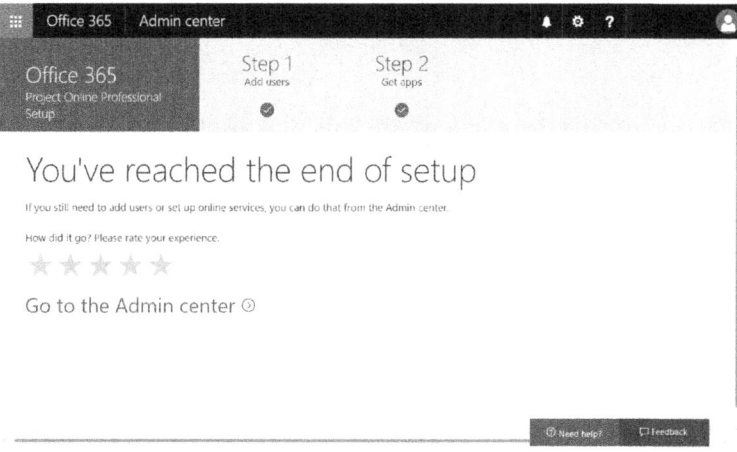

NOTE: *It may take a few minutes for the SharePoint admin center to be available.*

13. On the **Admin center** page, in the side menu, expand **Admin centers** and click **SharePoint**. Note: This will open a new tab in the browser.

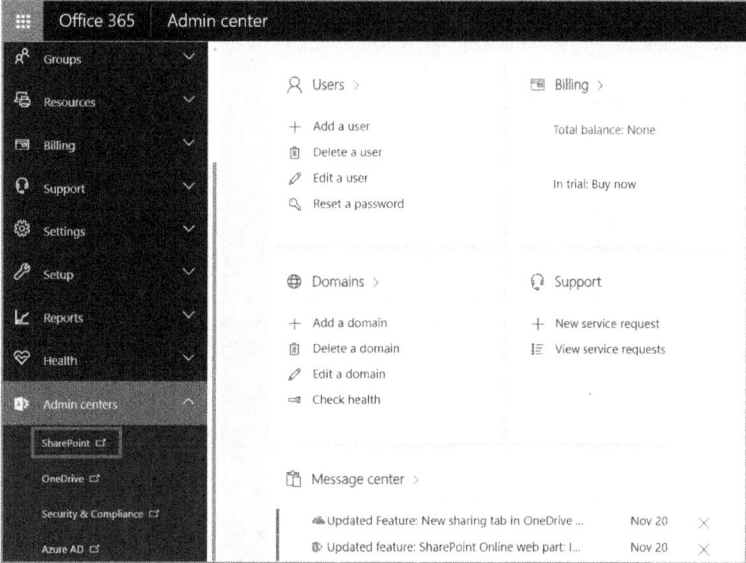

14. On the **SharePoint admin center** page, in the **Site Collections** list, click on the URL for Project Online: **https://<your domain url>/sites/pwa**.

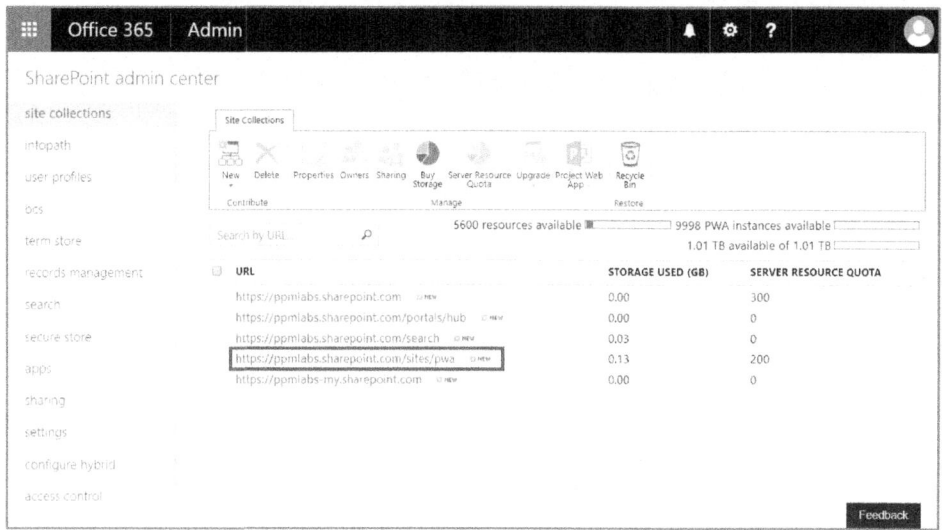

15. In the **site collection properties** window, click the link for the **Web Site Address**.

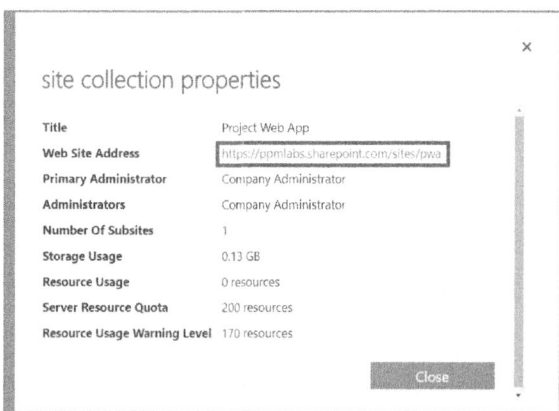

16. Project Web App is displayed

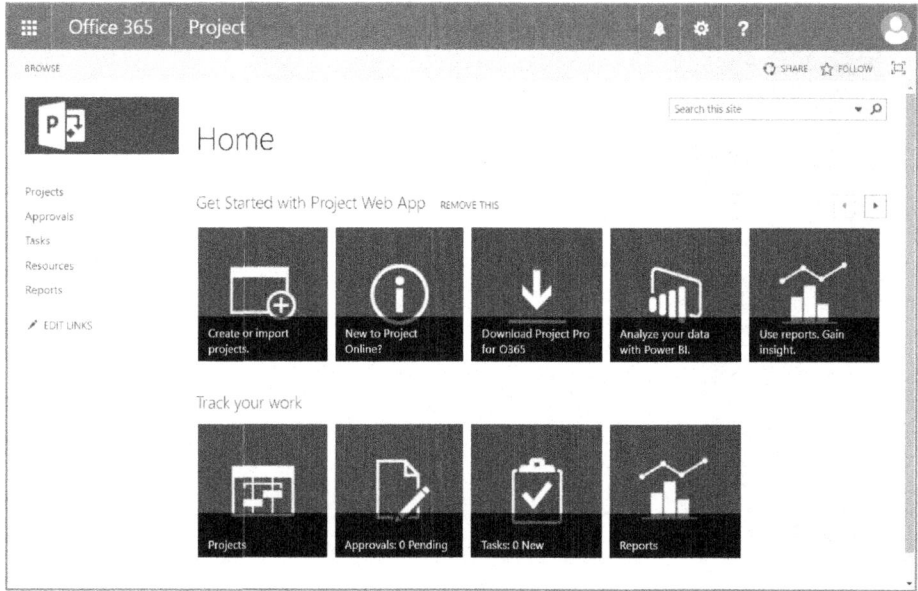

Adding a Service to an Existing Subscription

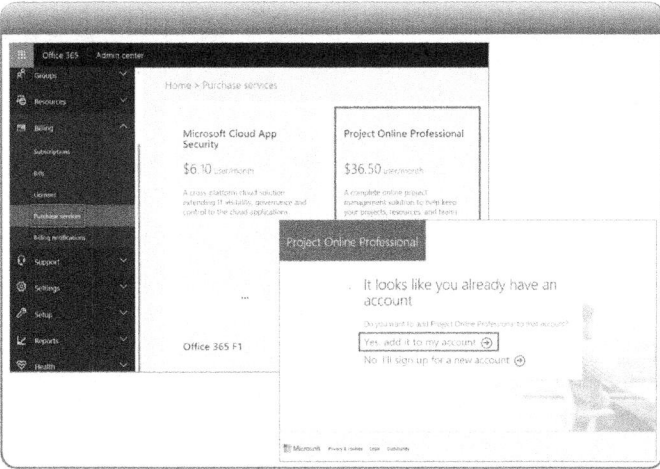

If you already have an existing subscription of Office 365, you can add Project Online in many ways.

Office 365 Admin Center

1. From the **Office 365 Admin center** page, in the side menu, expand **Billing** and click **Purchase services**.

2. From the **Purchase services** page, select the subscription and click **Buy now**.

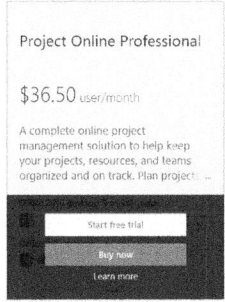

3. On the **Admin** page, select either **Pay monthly** or **Pay for a full year,** then add the number of users you want and click **Check out now**.

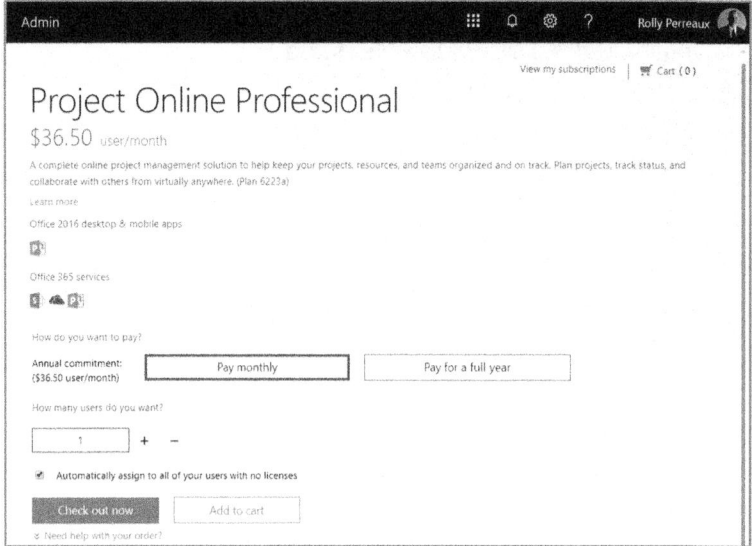

Microsoft Product Website

If you are already logged in to Office.com and you are attempting to purchase another subscription through a Microsoft product website, you may be given the opportunity to add a service to your subscription, as shown below:

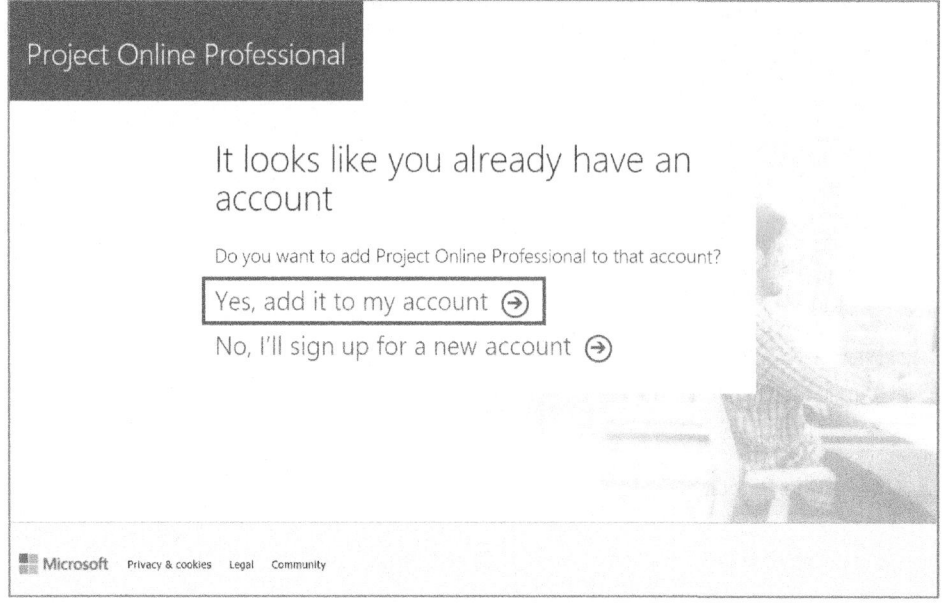

Practice: Creating a New Project Online Subscription

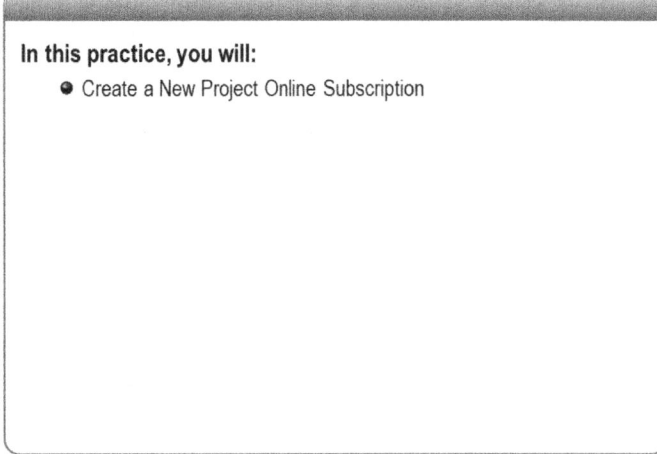

In this practice, you will:
- Create a New Project Online Subscription

Exercise 1: Create a New Project Online Subscription

In this exercise, you will create a Microsoft Project Online trial subscription.

NOTE: You will not need to use a credit card to create a trial subscription.

1. Launch your web browser and in the URL box, type https://www.microsoft.com/project

2. From the **Project** web page, click **See products & pricing**.

3. On the **Compare project management solutions** page, click **Try now** for either Project Online Professional.

4. On the **Welcome, let's get to know you** page, complete the form and click **Next**.

   NOTE: Use a valid email address as you will receive an email from Microsoft Office 365 with your account information, including User ID.

5. On the **Create your user ID** page, complete the following settings and click **Create my account**.

Setting	Perform the following:
Username	Type your first name
Yourcompany	Type a fictitious company name
Password	Type your password

6. On the **Save this info. You'll need it later** page, copy the information and click **You're ready to go…**

7. On the **Office 365** page, under the **Apps** section, click **Admin**.

8. On the **Admin center** page, in the **Welcome to the admin center** window, click **Next** and continue through a series of windows and then click **Finish.**

9. On the **Admin center** page, click **Go to setup**.

10. On the **Add new users** page, do not add any users as we will be doing this in the next practice, so click **Next**.

11. On the **Install your Office apps** page, do not install **Project Pro for Office 365**, as we will be doing this in Module 2, so click **Next**.

12. On the **You've reached the end of setup** page, click **Go to the Admin center**.

13. On the **Admin center** page, in the navigation menu, expand **Admin centers** and click **SharePoint**. (This will open a new tab in the browser)

    NOTE: It may take a few minutes for the SharePoint admin center to be available.

14. On the **SharePoint admin center** page, in the **Site Collections** list, click on the URL for Project Online: **https://<your domain url>/sites/pwa**.

15. In the **site collection properties** window, click the link for the **Web Site Address**.

16. The Project Web App site is displayed.

Lesson 2: Working with Office 365 Admin Center

- Adding New Users to Office 365
- Assigning Product Licenses to Users
- Creating Groups in Office 365
- Assigning Users to Groups

Once Project Online has been deployed, you will need to add Office 365 users and groups allow users to work with Project Online. In this lesson you will learn how to add new users to Office 365 and assign product licenses to users. You will also learn how to create groups and assign users to those groups.

Objectives

After completing this lesson, you will be able to:

- Add New Users to Office 365
- Assign Product Licenses to Users
- Create Groups in Office 365
- Assign Users to Groups

Adding New Users to Office 365

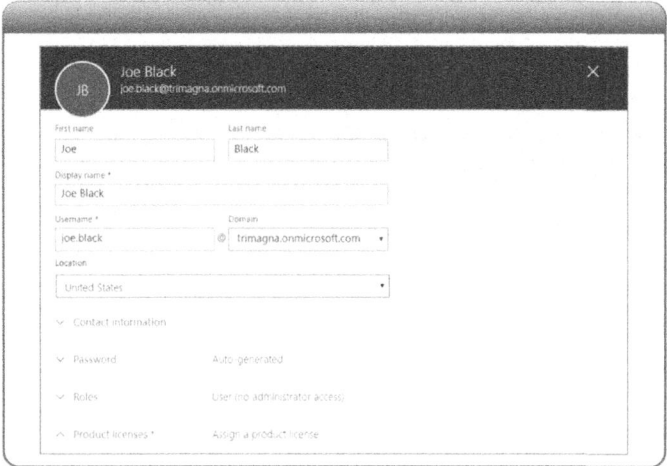

All users each need a user account before they can sign in and access Project Online. The easiest way to add user accounts is to add them one at a time in the Office 365 admin center.

Adding a User in the Office 365 Admin Center

1. In the **Office 365 admin center**, choose **Users** > **Active users** and then click **Add a user**.
2. Complete the following information for the user account:

Name	Complete first name, last name, display name, and user name.
Domain	For example, if the user's username is Joe, and his domain is contoso.com, he'll sign in to Office 365 by typing joe@contoso.com
Contact information	Expand to fill in a mobile phone number, address, and so on.
Password	Use the auto-generated password or expand to specify a strong password for the user.
Roles	Expand if you need to make this user an admin.
Product licenses	Expand this section and select the appropriate license. If you don't have any licenses available, you can still add a user and purchase additional licenses.

3. When you are finished, click **Add**.

Adding Multiple Users in the Office 365 Admin Center

4. In the **Office 365 admin center**, choose **Users** > **Active users**.
5. In the **Admin center** choose **Users** and then **Active users**.
6. In the **More** drop-down, choose **Import multiple users**.

Module 1: Working with Microsoft Project Online   1-15

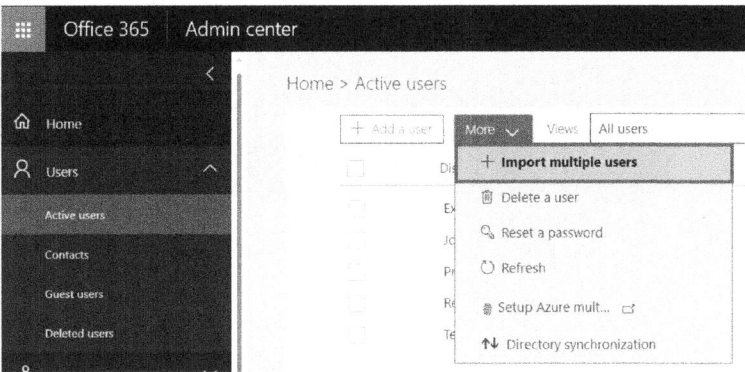

7. On the **Import multiple users** pane, you can optionally download a sample CSV file with or without sample data filled in.

   Your spreadsheet needs to include the exact same column headings as the following fields: *User Name,First Name,Last Name,Display Name,Job Title,Department,Office Number,Office Phone,Mobile Phone,Fax,Address,City,State or Province,ZIP or Postal Code,Country or Region*

   *NOTE: The **User name** and **Display name** fields are required fields.*

8. Enter a file path into the box, or choose **Browse** to browse to the CSV file location, then choose **Verify**.

9. On the **Set user options** dialog box, you can set the sign-in status and choose the product license that will be assigned to all users.

10. On the **View your result** dialog box, you can choose to send the results to either yourself or other users (passwords will be in plain text) and you can see how many users were created, and if you need to purchase more licenses to assign to some of the new users.

Assigning Product Licenses to Users

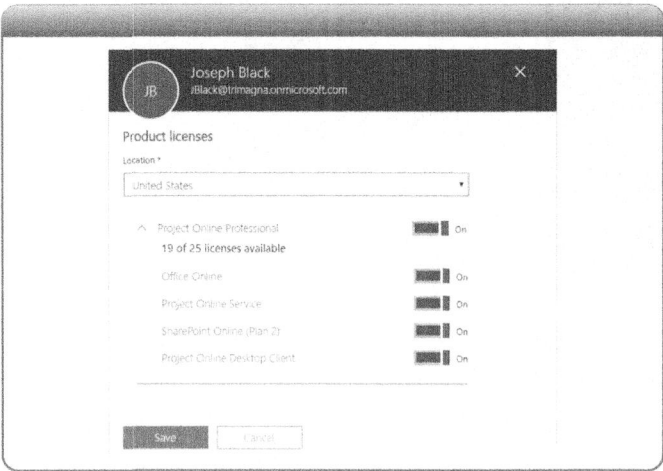

In order to use Project Online and the subsequent services a user must be assigned a product license. Then each user can install Project Online Desktop Client on up to five computers. Each installation is activated and kept activated automatically by cloud-based services associated with Office 365. This means you don't have to keep track of product keys. It also means you don't have to figure out how to use other activation methods such as Key Management Service (KMS) or Multiple Activation Key (MAK). All you must do is make sure you purchase enough licenses, keep your Office 365 subscription current, and make sure your users connect to Office Licensing Service via the Internet at least once every 30 days.

1. In the **Office 365 admin center**, go to the **Active users** page, or choose **Users > Active users**.

2. Select the box next to the name of the user to whom you want to assign a license.

3. On the right, in the **Product licenses** row, choose **Edit**.

4. In the **Product licenses** pane, switch the toggle to the **On** position for the license that you want to assign to this user.

   NOTE: By default, all services associated with that license are automatically assigned to the user. To limit which services are available to the user, switch the toggles to the Off position to restrict user services.

5. At the bottom of the **Product licenses** pane, choose **Save > Close > Close**.

Creating Groups in Office 365

> **There are 4 types of Groups**
> - **Office 365 Groups** are a great way for teams to collaborate by giving them a group email and a shared workspace for conversations, files, and calendar events.
> - **Distribution lists** send email to all members of the list. You can even allow people outside your organization send email to a list.
> - **Mail enabled security groups** can be used to control access to OneDrive and SharePoint as well as to send email to all members of the list.
> - **Security groups** control access to OneDrive and <u>SharePoint</u>

Groups in Office 365 let you choose a set of people that you wish to collaborate with and easily set up a collection of resources for those people to share.

There are four types of groups:

Office 365 groups are a great way for teams to collaborate by giving them a group email and a shared workspace for conversations, files, and calendar events.

Distribution lists send email to all members of the list. You can even allow people outside your organization to send email to a list.

Mail enabled security groups can be used to control access to OneDrive and SharePoint as well as to send email to all members of the list.

Security groups control access to OneDrive and SharePoint and are used for Mobile Device Management for Office 365. Using security groups in Project Online is strongly recommended

Private vs Public Office 365 Groups

When creating an Office 365 group, you'll need to decide if you want the group to be private or public. Content in a public group can be seen by anybody in your organization, and anybody in your organization is able to join the group. Content in a private group can only be seen by only the members of the group and people who want to join a private group must be approved by a group owner.

Neither public groups nor private groups can be seen or accessed by people outside of your organization unless those people have been specifically invited as guests.

Assigning Office 365 Groups to Users

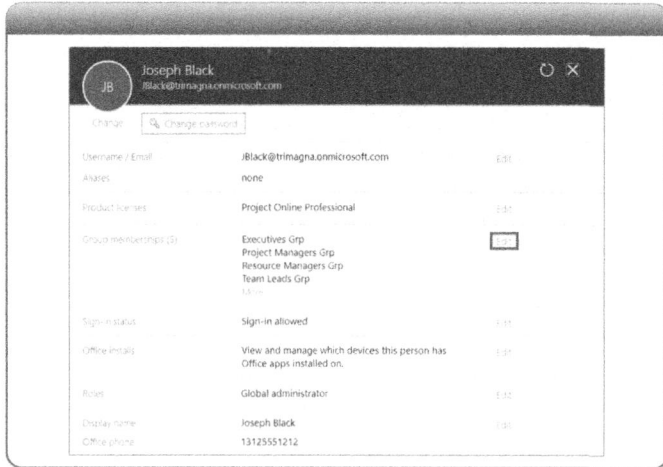

Once the Office 365 users and groups have been created, you will need to assign them appropriately. You can do this in two ways:

To Add a Group to a User

1. From the **Office 365 admin center** page, in the navigation pane, choose **Users > Active users**.
2. On the **Home > Active users** page, click on the **Display Name**.
3. In the **details** pane, under **Group memberships** section, click **Edit**.
4. In the **details** pane, click **+ Add memberships**.
5. Under the **Groups** listing, select the checkboxes for the groups to add and click **Save**.
6. In the **details** pane, click **Close** three times.

To Add a User to a Group

1. From the **Office 365 admin center** page, in the navigation pane, choose **Groups > Groups**.
2. On the **Home > Group** page, click on a **Group**.
3. In the details pane, next to **Members**, click **Edit**.
4. Search for or select the name of the member you want to add.
5. Click **Save**.

Practice: Working with Office 365 Admin Center

> **In this practice, you will:**
> - Add Office 365 Users
> - Assign Product Licenses to Users
> - Create Office 365 Groups
> - Assign Office 365 Groups to Users

Exercise 1: Adding Office 365 Users

In this exercise, you will manually add a new user to the Office 365, you will then import multiple users by uploading a CSV file.

1. Switch to the **Office 365 Admin Center** web site at: https://portal.office.com/adminportal.
2. From the **Home** page, in the navigation menu, expand **Users** and click **Active Users**.
3. On the **Home > Active users** page, click **+ Add a user**.
4. In the **New user** pane, complete the form with the following settings and click **Add**.

Setting	Perform the following:
First name	Type **Karla**
Last name	Type **Carter**
Display name	Should be automatically listed as **Karla Carter**
Username	Type **KCarter**
Location	Leave as default location.
Expand Contact Information	Job Title: Type **Director of Programs** Department: Type **Programs** Office: Type **301** Office Phone: Type **555-1212 x.301**
Expand Password	Select **Let me create the password**; In the **Password** box, type **Pa$$w0rd**; Clear checkbox, **Make this user change their password when they first sign in**.
Expand Product licenses	Select **Create user without product license**.

NOTE: We will assign product licenses in the next exercise.

5. In the **User was added** pane, clear checkbox, **Send password in email** and click **Close**.
6. On the **Home > Active users** page, click **More** and click **Import multiple users**.
7. In the **Import multiple users** pane, click **Browse**.
8. In the **Choose File to Upload** dialog box, navigate to the location where the course files were saved, select **Import_Users.csv** and click **Open**.
9. In the **Import multiple users** pane, click **Verify**.

    If you receive the message: "**Your file looks good. Click or tap Next.**" then click **Next**.

10. In the **Import multiple users** pane, under **Product licenses** section, select **Create user without product license** and click **Next**.

NOTE: We will assign product licenses in the next exercise.

11. In the **Import multiple users** pane, clear the check box **Email the results files to these people** and click **Close without sending**.
12. On the **Home > Active users** page, select all check boxes (except your account).

NOTE: There should be 24 users selected

13. In the **Bulk actions** pane, click **Reset passwords**.
14. In the **Reset passwords** pane, select **Let me create the password**, in the **Password** box, type **Pa$$w0rd**, then clear checkbox, **Make this user change their password when they first sign in** and click **Reset**, as shown below:

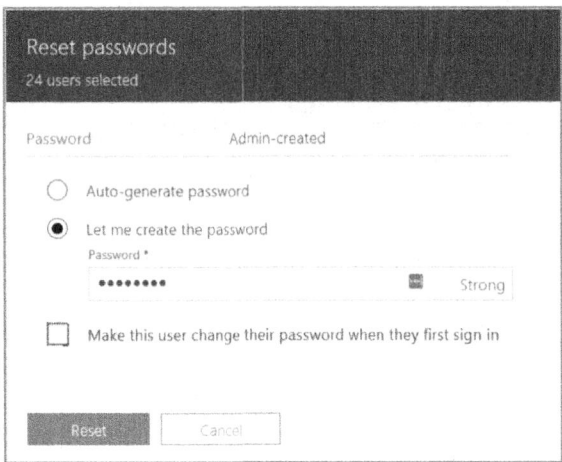

15. In the **Reset passwords** pane, clear check box **Send password in email** and click **Close**.

Exercise 2: Assigning Product Licenses to Users

In this exercise, you will assign a Microsoft Project Online license to users using bulk action.

1. In the **Bulk actions** pane, click **Edit product licenses**.
2. In the **Assign products** pane, select **Add to existing product license assignments** and click **Next**.

3. In the **Add to existing products** pane, turn **Project Online Professional** to **On** and accept the default settings and click **Add**.

4. In the **Add to existing products** pane, click **Close**.

5. In the **Bulk Actions** pane, click **Close**.

Exercise 3: Creating Security Groups

In this exercise, you will create seven new security groups.

1. From the **Home > Active users** page, in the navigation menu, expand **Groups** and click **Groups**.

2. On the **Home > Groups** page, click + **Add a group**.

3. In the **New Group** pane, complete the form with the following settings and click **Add**.

Setting	Perform the following:
Type	Select **Security group**
Name	Type **Project Managers Grp**

4. In the **Group was added** pane, click **Close**.

5. Repeats Steps 2-4 for the following groups:

   Resource Managers Grp

   Business Analysts Grp

   Executives Grp

   PMO Managers Grp

   Team Leads Grp

   Team Members Grp

Exercise 4: Assigning Users to Security Groups

In this exercise, you will assign users to the newly created security groups.

1. On the **Home > Groups** page, select the check box for **Business Analysts Grp**.

2. In the **Business Analysts Grp** pane, complete the form with the following settings.

Setting	Perform the following:
Owners	Click **Edit**.
	Click + **Add owners**
	Select **<your username>** and click **Save**.
Members	Click **Edit**.
	Click + **Add members**
	Select **Michael Schmitz** and click **Save**.

3. In the **Business Analysts Grp** pane, click **Close** 3 times.
4. Repeat Steps 1-3 for the following security groups:

Setting	Perform the following:
Executives Grp	Owners: **<your username>** Members: **Jeanette Frazier, Jennifer Herman, Karla Carter, Mac Ellington, Nick Portlock, Robert Cain, Sigurdur Haraldsson** and **Ted Malone**
PMO Managers Grp	Owners: **<your username>** Members: **Marco Shaw**
Project Managers Grp	Owners: **<your username>** Members: **Beth Quinlan** and **Karla Carter**
Resource Managers Grp	Owners: **<your username>** Members: **Brian Rieck**
Team Leads Grp	Owners: **<your username>** Members: **Eric Denekamp**
Team Members Grp	Owners: **<your username>** Members: Select all 25 users (not groups)

Summary

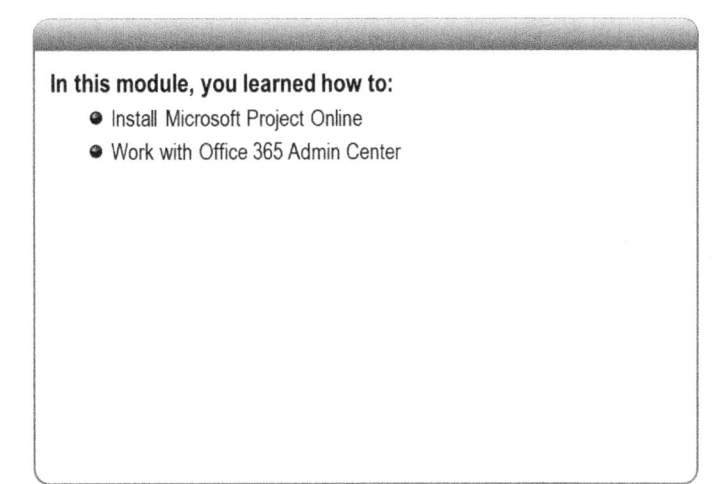

> **In this module, you learned how to:**
> - Install Microsoft Project Online
> - Work with Office 365 Admin Center

In this module, you learned the features and installed Microsoft Project Online. You also learned about the Office 365 Admin Center and how to manage Office 365 users and groups as they are the foundation for a successful portfolio management (PPM) solution with Project Online.

Objectives

After completing this module, you learned how to:

- Install Microsoft Project Online
- Work with Office 365 Admin Center

This page is intentionally left blank

PMO Logistics

Module 2: Managing Project Online Security

Contents

Module Overview 1
Lesson 1: Overview of Project Online Security 2
    Understanding Permission Modes 3
    Security Workflow – SharePoint Permission Mode 4
    Security Workflow – Project Server Permission Mode 5
    Project Online Security Entities 6
Lesson 2: SharePoint Security Permissions 7
    Default SharePoint Groups 8
    Adding Users to SharePoint Groups to Access PWA 9
    Switching Permission Mode 10
    Practice: Working with SharePoint Security Permissions 12
Lesson 3: Project Online Security Permissions 15
    Project Online Security Permissions 16
    Global Permissions 17
    Category Permissions 18
    Category Dynamics Rules 19
    Security Templates 21
    Practice: Creating Security Templates 22
Lesson 4: Creating Project Online Security Principals 25
    Creating Project Online Users 26
    Understanding Project Online Groups 29
    Creating Project Online Groups 31
    Understanding Project Online Categories 32
    Creating Project Online Categories 33
    Practice: Working with Project Online Security Principals 35
Summary 39

EXCLUSIVELY PUBLISHED BY

PMO Logistics
679 Roberta Avenue
Winnipeg, Manitoba, Canada R2K 0K9

Copyright © 2017 by Roland Perreaux

All rights reserved. No part of the contents of this document may be reproduced or transmitted in any form or by any means without written permission of the publisher.

Information in this document, including URL and other Internet Web site references, is subject to change without notice. Unless otherwise noted, the example companies, organizations, products, domain names, e-mail addresses, logos, people, places, and events depicted herein are fictitious, and no association with any real company, organization, product, domain name, e-mail address, logo, person, place, or event is intended or should be inferred. Complying with all applicable copyright laws is the responsibility of the user. Without limiting the rights under copyright, no part of this document may be reproduced, stored in or introduced into a retrieval system, or transmitted in any form or by any means (electronic, mechanical, photocopying, recording, or otherwise), or for any purpose, without the express written permission of PMO Logistics Inc.

The names of manufacturers, products, or URLs are provided for informational purposes only and PMO Logistics makes no representations and warranties, either expressed, implied, or statutory, regarding these manufacturers or the use of the products with any Microsoft technologies. The inclusion of a manufacturer or product does not imply endorsement of Microsoft of the manufacturer or product. Links are provided to third party sites. Such sites are not under the control of PMO Logistics and PMO Logistics is not responsible for the contents of any linked site or any link contained in a linked site, or any changes or updates to such sites. PMO Logistics is not responsible for webcasting or any other form of transmission received from any linked site. PMO Logistics is providing these links to you only as a convenience, and the inclusion of any link does not imply endorsement of PMO Logistics of the site or the products contained therein.

PMO Logistics may have patents, patent applications, trademarks, copyrights, or other intellectual property rights covering subject matter in this document. Except as expressly provided in any written license agreement from PMO Logistics, the furnishing of this document does not give you any license to these patents, trademarks, copyrights, or other intellectual property.

PMO Logistics, Professional Training Series, Upgrading Skills Series, TriMagna Corporation and TriMagna Corporation logo are either registered trademarks or trademarks of PMO Logistics Inc. in Canada, the United States and/or other countries.

Microsoft, Active Directory, Internet Explorer, Outlook, Project Server, SharePoint, SQL Server, Visual Studio, Windows and Windows Server are either registered trademarks or trademarks of Microsoft Corporation in the United States and/or other countries.

All other trademarks are property of their respective owners.

Author:	Rolly Perreaux, PMP, MCSE, MCT
Publisher:	PMO Logistics
Developmental Editor:	Heather Perreaux
Cover Graphic Design:	Andrea Ardiles
Technical Testing:	Underground Studioworks

Post-Publication:
Errata List Contributors:

Module Overview

- Overview of Project Online Security
- SharePoint Security Permissions
- Project Online Security Permissions
- Creating Project Online Security Entities

Project Online enables authenticated and authorized users the ability to create, manage, publish, update and view project information. This information is very valuable and needs to be secured so that only individuals or groups of individuals can work or view the data.

In order to secure the data, security permissions must be granted depending on the individual's role, whether the individual is employed by the organization (internal) or as an external stakeholder. For each user to perform their work, their role will require different permissions.

Objectives

After completing this module, you will be able to:

- Understand how Project Online security works
- Work in SharePoint Permission Mode
- Work in Project Server Permission Mode
- Create Project Online Security Entities:
    - Project Online users
    - Project Online groups
    - Project Online categories

Lesson 1: Overview of Project Online Security

- Understanding Permission Modes
- SharePoint Permission Mode
- Project Permission Mode
- Project Online Security Workflow
- Project Online Security Entities

The Project Online security model now offers two security modes for controlling the kind of access that users have to projects and project sites and.

In this lesson, you will learn about SharePoint Permission Mode and Project Permission Mode. You will also learn how security workflow and security entities work in Project Online.

Objectives

After completing this lesson, you will be able to:

- Understand the following Permission Modes:
    - SharePoint Permission Mode
    - Project Permission Mode
- Explain how the Project Online security workflow work
- Describe the Project Online security entities

Understanding Permission Modes

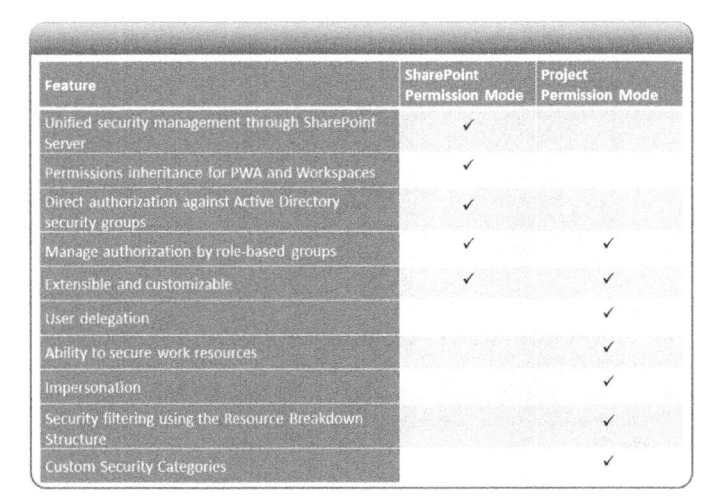

Project Online offers two permission modes for controlling user access to projects and project sites:

SharePoint Permission Mode

In this mode, a special set of SharePoint security groups are created in sites associated with Project Online. These groups are used to grant users varying levels of access to projects and Project Online functionality.

Project Permission Mode

In this mode, Project Online provisions a set of customizable security groups and other functionality that is distinct from SharePoint groups. This is the same security mode that was available in Project Server 2010 and earlier.

Feature	SharePoint Permission Mode	Project Permission Mode
Unified security management through SharePoint Server	✓	
Permissions inheritance for PWA and Project Sites	✓	
Direct authorization against Active Directory security groups	✓	
Manage authorization by role-based groups	✓	✓
Extensible and customizable	✓	✓
User delegation		✓
Ability to secure work resources		✓
Impersonation		✓
Security filtering using the Resource Breakdown Structure		✓
Custom Security Categories		✓

Security Workflow – SharePoint Permission Mode

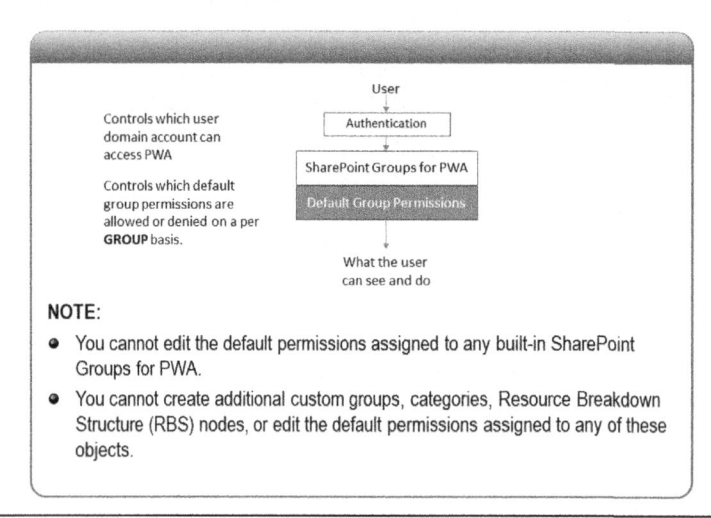

> Controls which user domain account can access PWA
>
> Controls which default group permissions are allowed or denied on a per **GROUP** basis.
>
> User → Authentication → SharePoint Groups for PWA → Default Group Permissions → What the user can see and do
>
> **NOTE:**
> - You cannot edit the default permissions assigned to any built-in SharePoint Groups for PWA.
> - You cannot create additional custom groups, categories, Resource Breakdown Structure (RBS) nodes, or edit the default permissions assigned to any of these objects.

In SharePoint Permission Mode, users are directly assigned to a built-in SharePoint Group for PWA or by being members of one or more security groups. When a user attempts to access PWA, the security workflow model goes through a two-stage process.

1. **Authentication Process** – In this stage, the PWA administrator controls which Office 365 accounts are added to the SharePoint Groups for PWA to access Project Web App.

2. **SharePoint Groups for PWA** – In this stage, default permissions (non-editable) are assigned to the built-in groups and are reviewed to determine if the group is allowed preconfigured global and category permissions.

NOTE: *In SharePoint Permission Mode, you cannot edit the default permissions assigned to any of these SharePoint groups. You also cannot create additional custom groups, categories, Resource Breakdown Structure (RBS) nodes, or edit the default permissions assigned to any of these objects.*

Change to Project Permission Mode if you need scalable and control management of user permissions.

Security Workflow – Project Server Permission Mode

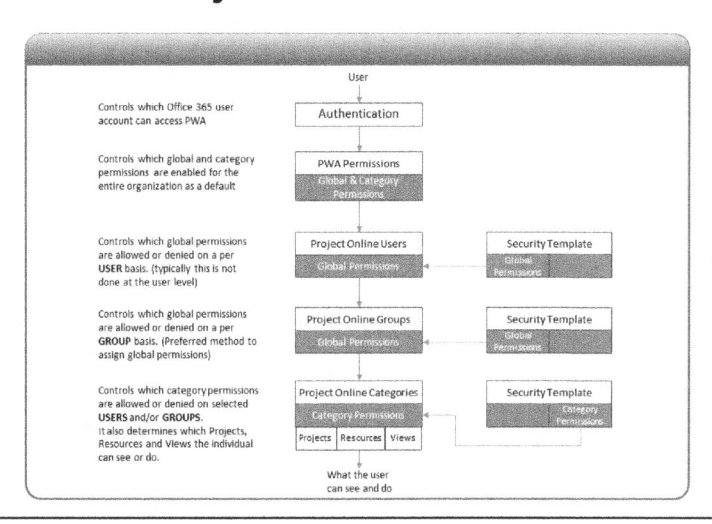

In Project Online, security principals are assigned permissions directly or by applying a security template to access security objects in categories. When a user attempts to access PWA, the security workflow model goes through a five-stage process.

1. **Authentication Process** – In this stage, the PWA administrator controls which Office 365 user accounts will access PWA.

2. **PWA Permissions** - In this stage, the PWA administrator controls which global and category permissions are enabled in Project Online for the entire organization. The administrator essentially either turns on or turns off Project Online permissions and features for all users.

3. **Project Online Users** - In this stage, global permissions assigned to the user is reviewed to determine if the user is allowed or denied specific global permissions. It also determines the group membership list that the user belongs to and if the user is directly assigned to any categories.

4. **Project Online Groups** - In this stage, global permissions assigned to the group is reviewed to determine if the group is allowed or denied specific global permissions and if the group is directly assigned to any categories.

5. **Project Online Categories** – In this final stage, all the individual security entities come together. The Project Online administrator controls which category permissions are allowed or denied based on the selected users and/or groups on an individual basis. It also determines which projects, resources and views are available for the users and groups.

The Project Online security model is further divided into three areas:

- Security objects
- Security principals
- Security templates

Project Online Security Entities

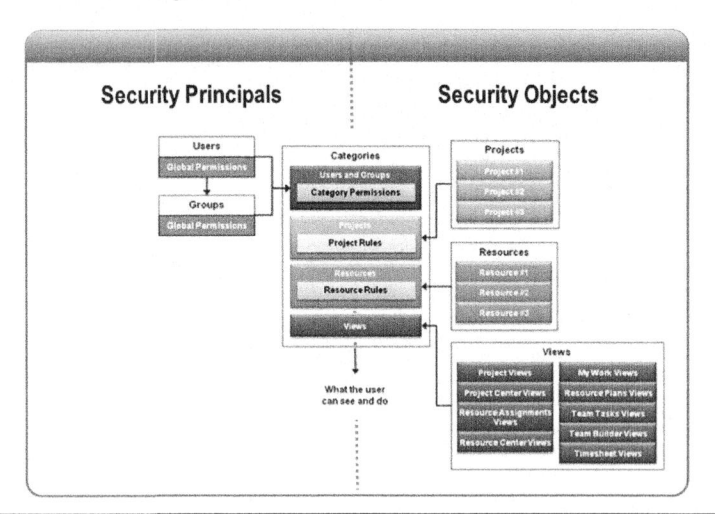

Security Objects

A security object is an entity that can be secured with permissions to be accessed by users and/or groups (security principals). These entities can have actions carried out on them or contain data. In Project Online, security objects consist of the following:

Projects – Define the sets of data fields that can be displayed for the collections of projects. Project data is very extensive and can include tasks, assignments, milestones and costs, to name a few.

Resources – A resource is traditionally defined as any of the people, equipment and materials used to complete tasks in a project.

Views – A Project Online view is a collection of table fields, formats and filters that are displayed for the selected projects, assignments, and resources in a category. Think of a view as a data display to what a user can see and do. To provide a user or group access to a view, you will need to assign the view in a category, and then grant category permissions to the user or group.

Categories – A category is considered a security object meaning the category can be secured with permissions to be accessed by users or groups (security principals). Categories bind all security principals and objects together.

IMPORTANT: Categories are a unique Project Online security entity as they are both a security object and a security principal, because categories are secured by category permissions, but they also grant user's or group's permission to view security objects.

Security Principals

A security principal is an entity that is granted permissions on security objects. Security principals are essentially individuals or groups of individuals that require access to the Project Online data and that data is contained in security objects. These principals need to have rights and permissions to access, use and view the data in the security objects. Project Online security principals are **Users**, **Groups** and **Categories**.

Lesson 2: SharePoint Security Permissions

- Default SharePoint Groups for PWA
- Adding Users to SharePoint Groups for PWA
- Switching Security Permission Mode

In this lesson, you will learn about SharePoint security permissions

Objectives

After completing this lesson, you will be able to:

- Know the default SharePoint Groups for PWA
- Add users to SharePoint Groups for PWA
- Switch security permission mode

Default SharePoint Groups

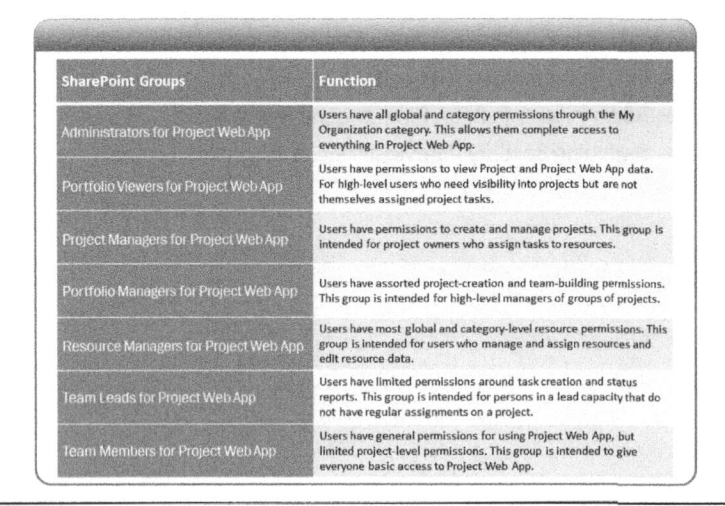

SharePoint Groups	Function
Administrators for Project Web App	Users have all global and category permissions through the My Organization category. This allows them complete access to everything in Project Web App.
Portfolio Viewers for Project Web App	Users have permissions to view Project and Project Web App data. For high-level users who need visibility into projects but are not themselves assigned project tasks.
Project Managers for Project Web App	Users have permissions to create and manage projects. This group is intended for project owners who assign tasks to resources.
Portfolio Managers for Project Web App	Users have assorted project-creation and team-building permissions. This group is intended for high-level managers of groups of projects.
Resource Managers for Project Web App	Users have most global and category-level resource permissions. This group is intended for users who manage and assign resources and edit resource data.
Team Leads for Project Web App	Users have limited permissions around task creation and status reports. This group is intended for persons in a lead capacity that do not have regular assignments on a project.
Team Members for Project Web App	Users have general permissions for using Project Web App, but limited project-level permissions. This group is intended to give everyone basic access to Project Web App.

When using SharePoint Permissions Mode, Project Online creates the following SharePoint Groups for PWA that directly correspond to the default security groups that can be found in Project Permission Mode.

Administrators for Project Web App – Users have all global and category permissions through the My Organization category. This allows users complete access to everything in Project Web App.

Portfolio Viewers for Project Web App – Users have permissions to view Project and PWA data. This group is intended for high-level users who need visibility into projects but are not themselves assigned project tasks.

Project Managers for Project Web App – Users have permissions to create and manage projects. This group is intended for project owners who assign tasks to resources.

Portfolio Managers for Project Web App – Users have assorted project-creation and team-building permissions. This group is intended for high-level managers of groups of projects.

Resource Managers for Project Web App – Users have most global and category-level resource permissions. This group is intended for users who manage and assign resources and edit resource data.

Team Leads for Project Web App – Users have limited permissions around task creation and status reports. This group is intended for persons in a lead capacity that do not have regular assignments on a project.

Team Members for Project Web App – Users have general permissions for using PWA, but limited project-level permissions. This group is intended to give everyone basic access to PWA.

Adding Users to SharePoint Groups to Access PWA

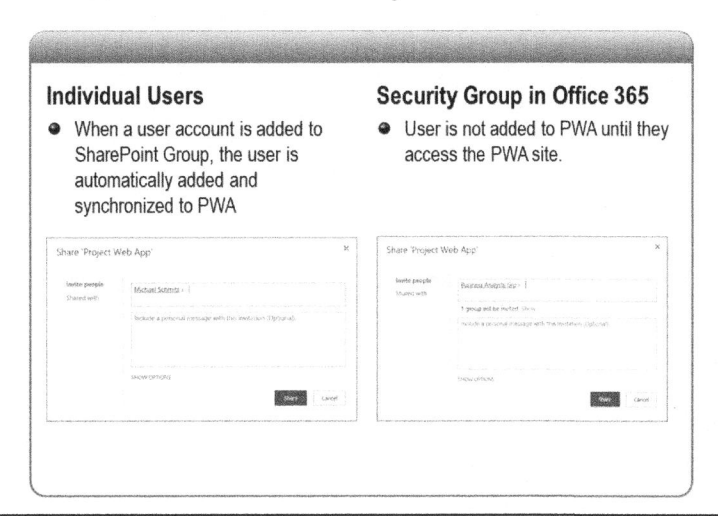

There are two methods of adding users to SharePoint Groups for PWA:

Individual Users

When you add individual users to one of the SharePoint groups, that user is synchronized to PWA automatically. User synchronization runs on a SharePoint timer job, by default every ten minutes.

Security Groups in Office 365

When you add a security group to one of the PWA-specific SharePoint security groups, the users are not automatically added to the list of users in PWA. Each user is individually added to PWA the first time she or he accesses the PWA site.

Because users in security groups do not appear on the list of PWA resources until they have accessed the PWA site, it is recommended that you configure Active Directory (Office 365) synchronization in PWA to pre-populate your resource list. This allows you to have a complete resource list and to assign work to resources before they have accessed the PWA site.

NOTE: *You can use either of these methods for each group.*

Switching Permission Mode

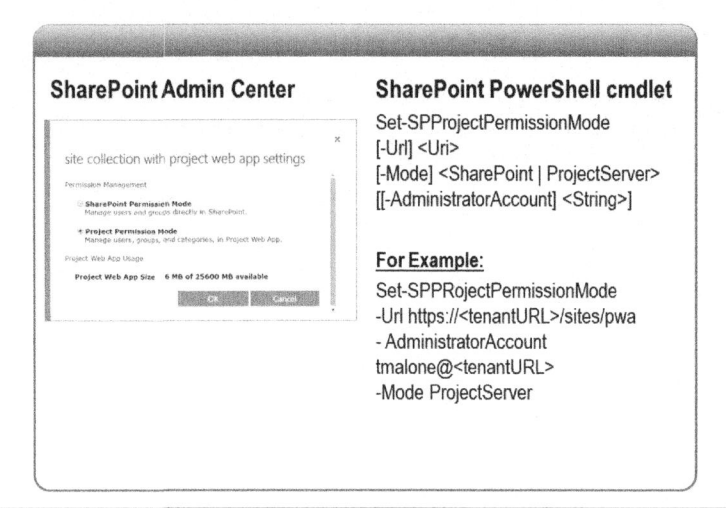

SharePoint Admin Center

site collection with project web app settings

Permission Management
- SharePoint Permission Mode
  Manage users and groups directly in SharePoint.
- Project Permission Mode
  Manage users, groups, and categories, in Project Web App.

Project Web App Usage
Project Web App Size   6 MB of 25600 MB available

OK   Cancel

SharePoint PowerShell cmdlet

Set-SPProjectPermissionMode
[-Url] <Uri>
[-Mode] <SharePoint | ProjectServer>
[[-AdministratorAccount] <String>]

For Example:
Set-SPPRojectPermissionMode
-Url https://<tenantURL>/sites/pwa
- AdministratorAccount
tmalone@<tenantURL>
-Mode ProjectServer

As previously stated, Project Online offers two permission modes for controlling user access to projects and project sites. By default, all new PWA instances use SharePoint permission mode. However, you can switch permission mode using one of the following methods:

SharePoint Admin Center

1. Access the Office 365 Admin Center and in the navigation pane, click **Admin centers** and click **SharePoint**.

2. On the **SharePoint admin center** page, select the URL of the PWA tenant and on the **Site Collections** ribbon, click **Project Web App** and then click **Settings**.

3. In the **site collection with project web app settings** window, select either **SharePoint Permission Mode** or **Project Permission Mode** and click **OK**.

SharePoint PowerShell Cmdlet

Use the **Set-SPProjectPermissionModeWindows** PowerShell cmdlet to change the permission mode:

```
Set-SPProjectPermissionMode [-Url] <Uri> [-Mode] <SharePoint |
ProjectServer | UninitializedSharePoint |
UninitializedProjectServer> [[-AdministratorAccount] <String>]
```

Parameter	Required	Description
Url	Required	Specifies the URL of the PWA instance for which the permission mode is to be changed. The type must be a valid URL, in the form of: *http://<ServerName>/<PWAName>*.

Mode	Required	Specifies the mode into which the instance should be changed. The type must be a valid permission mode, in the form of: **SharePoint** or **ProjectServer**.
AdministratorAccount	Optional	The name of the account to be added as a PWA administrator.

WARNING: Switching between SharePoint Permission Mode and Project Permission Mode deletes all security-related settings.

Therefore, if you switch from SharePoint Permission Mode to classic Project (Server) Permission Mode, you must manually configure your security permissions structure in Project Online. Switching from Project Permission Mode back to SharePoint Permission Mode deletes your security permissions information from Project Online.

Practice: Working with SharePoint Security Permissions

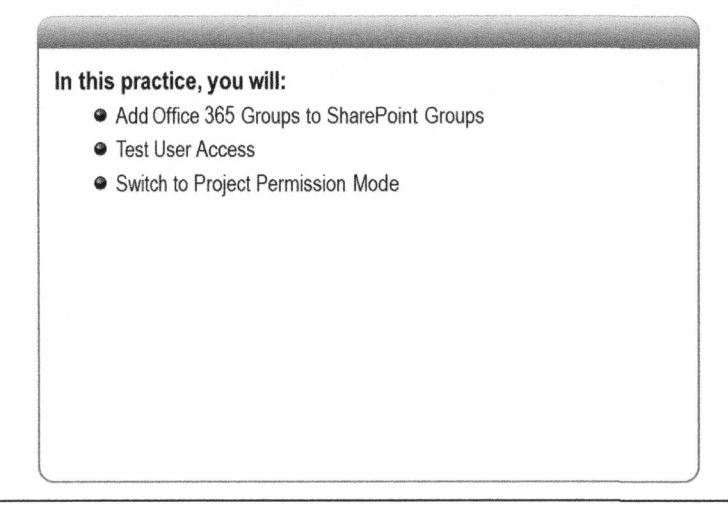

In this practice, you will:
- Add Office 365 Groups to SharePoint Groups
- Test User Access
- Switch to Project Permission Mode

In this practice, you will add security groups in Office 365 to the built-in SharePoint groups for PWA. You will then test to ensure that users can access the PWA site by using the Run As feature. You will also switch over to Project Permission Mode.

Exercise 1: Add Office 365 Groups to SharePoint Groups

In this exercise, you will add Office 365 users to SharePoint groups for PWA.

1. From your web browser, in the URL box, type **https://<companyname>.sharepoint.com/sites/pwa** and sign in as the Project Online Administrator.

2. On the **PWA Home** page, click the **Settings** menu (top right side) as shown below:

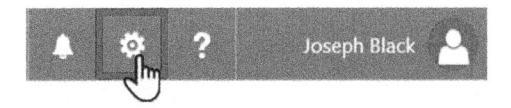

3. Then click **Site settings**.

4. On the **Site Settings** page, under **Users and Permissions** section, click **Site permissions**.

5. On the **Permissions** page, click **Portfolio Managers for Project Web App**.

6. On the **Portfolio Managers for Project Web App** page, click **New**.

7. On the **Share 'Project Web App'** window, on the **Enter names, email addresses...** box, type **Michael Schmitz** and click **Share**.

8. On the **Portfolio Managers for Project Web App** page, in the Quick Launch menu, under **Groups**, click **More…**.

9. On the **People and Groups** page, complete the configuration (using steps 5-8) for the following groups:

SharePoint Group	Add the following Office 365 User:
Portfolio Viewers for Project Web App	Mac Ellington
Project Managers for Project Web App	Beth Quinlan

Resource Managers for Project Web App	Brian Rieck
Team Leads for Project Web App	Eric Denekamp

Exercise 2: Test User Access

In this exercise, you will test that the user can access the Project Web App site.

1. From **Internet Explorer**, press **Ctrl + Shift + P** to start an **InPrivate** session.

2. In **Internet Explorer**, in the **Address** box, type **https://<companyname>.sharepoint.com/sites/pwa** and press **Enter**.

3. At the **Microsoft Sign in** page, type **BQuinlan@<companyname>.onmicrosoft.com** and click **Next**.

4. At the **Microsoft Enter password** page, type **Pa$$w0rd** and click **Sign in**.

5. On the personal menu, notice that **Beth Quinlan** is now listed and has access to PWA.

6. Click on Beth Quinlan's name (upper right) and click **Sign Out**.

7. Close **Internet Explorer**.

Exercise 3: Switch to Project Permission Mode

In this exercise, you will use Office 365 Admin center to switch the permission mode of Project Online.

1. From the **People and Groups** page, click on the **App Launcher** (top left) and click **Admin**, as shown below:

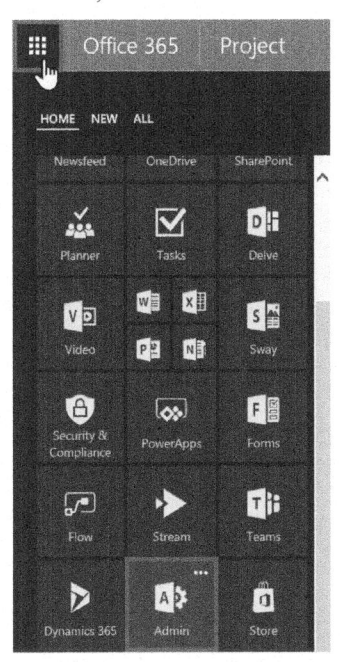

2. From the **Office 365 Admin center** page, in the navigation pane, click **Admin centers** and click **SharePoint**.

3. On the **SharePoint admin center** page, select the URL of the PWA tenant and on the **Site Collections** ribbon, click **Project Web App** and then click **Settings**, as shown below:

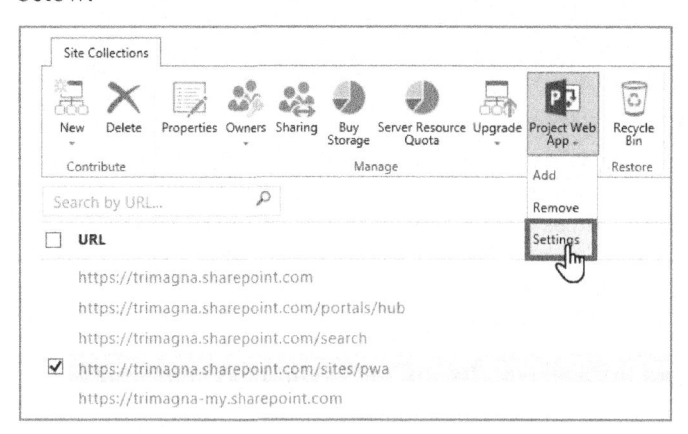

4. In the **site collection with project web app settings** window, select **Project Permission Mode** and click **OK**.

5. In **Internet Explorer**, in the **Address** box, type **https://<companyname>.sharepoint.com/sites/pwa** and press **Enter**.

6. On the **PWA Home** page, click **Settings** and click **PWA Settings**.

7. On the **PWA Settings** page, if the **Security** section is displayed, then the permission mode has switched to Project Permission Mode.

Lesson 3: Project Online Security Permissions

- **Project Online Security Permissions**
  - Global Permissions
  - Category Permissions
- **Category Dynamics Rules**
- **Security Templates**

In this lesson, you will learn about Project Online security permissions.

Objectives

After completing this lesson, you will be able to:

- Understand Project Online security permissions:
  - Global Permissions
  - Category Permissions
- Understand Category Dynamics Rules
- Create Project Online Security Templates

Project Online Security Permissions

- **Security permissions on Project Online are divided into two sets of groups:**
  - Global Permissions
  - Category Permissions
- **Permission Settings**
  - Allow – Explicitly allowed the permission
  - Deny – Explicitly denied the permission
  - None Selected – Implicitly denied the permission

Security permissions define the access to security objects for security principals (users and groups). They are the rules that determine the actions a user or group can perform on Project Online.

Security permissions on Project Online are divided into two sets of groups:

- Global Permissions
- Category Permissions

Security Permission Settings

Security permissions are applied on a server, user and group basis. This means that a user's actual permissions are a culmination of all the permissions enabled on the server, the user's permissions and the group's permission that the user is a member of. When setting these permissions, the Project Online administrator can explicitly set the permission to either Allow or Deny. They can also implicitly set permission to Not Allow by clearing the check box.

Permission Setting	Description
Deny	The user or group is explicitly denied the permission
Allow	The user or group is explicitly allowed the permission
None selected	If neither permission setting is selected, the user or group does not have access (Implicitly denied)

Deny always has the highest priority and is applied to a user everywhere that the permission affects. The key thing to remember when setting a permission to Deny, is that it overrides Allow.

For example, if Joe Smith was a member of a group that was explicitly assigned Deny for a particular permission, but his user account was explicitly assigned Allow for the same permission, his culminated permissions is Deny as it overrides everything else.

Global Permissions

3 Levels of Global Permissions:
- Server Level
- User Level
- Group Level

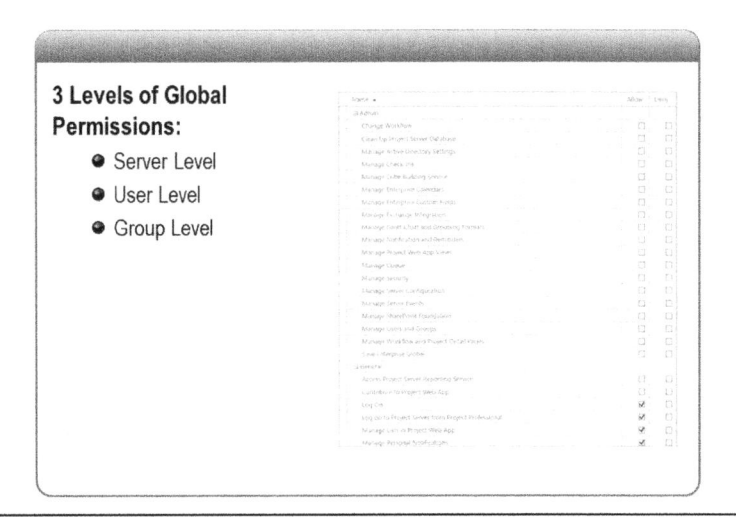

Global permissions can be set at three different levels:

Server Level

Global permissions at the Server Level are set at the PWA Permissions page. At the Server level, global and category permissions are either enabled or denied on Project Online. This means that if a Project Online administrator were to clear a check box for a particular feature or permission, it would be denied for all users, including the administrator. If the check box is enabled, then the feature or permission is available to use by users and groups. It should be noted that the lists of all features and permissions are sorted alphabetically and not divided between global and category permissions.

User Level

At the User level, the Project Online administrator can assign global permissions directly to users. However, unless you will be defining and using security templates to be implemented for user roles, applying global permissions at the user level should be done only for very special circumstances and for only a few users.

This is because assigning individual permissions for all users in the organization is a very time-consuming task for the Project Online administrator. It is also prone to assigning specific global permissions that the user should not have by error. A more administrative-friendly approach would be to assign global permissions to groups instead and then adding the users to the group. By default, there are global permissions pre-assigned to users.

Group Level

At the Group level, the Project Online administrator can assign the same global permissions directly to groups. The default groups already have global permissions assigned by default.

Category Permissions

- Deal with specific actions that a user and/or group can do with the projects and resources

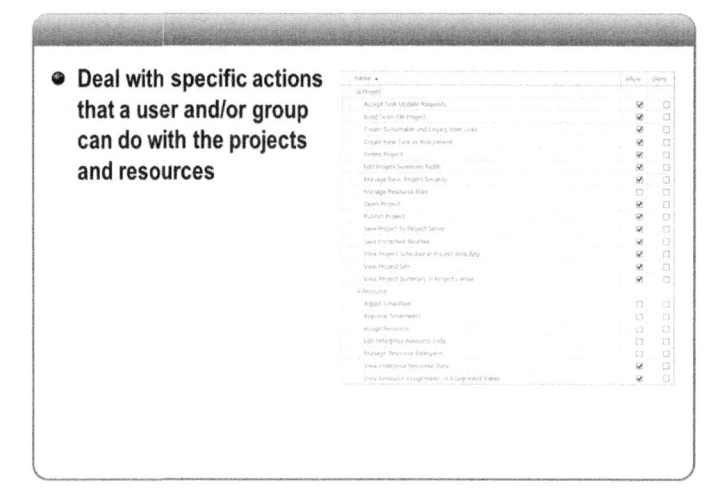

Category permissions are different than global permissions because they deal with specific actions that a user or group assigned to the category can do with the projects and resources listed in the category. The default categories already have category permissions assigned to the default groups. To view the listing of pre-assigned category permissions for the default categories, please review the appendix for more detailed information.

When a user or group is granted permission in a category, they are granted permission to all the assigned security objects in that category. For example, if a category contains two projects, granting a security principal the Open Project and Save Project to Project Server permissions in the category allows that user to open and save changes to both projects. Therefore, Project Online administrators must take care when assigning category permissions.

Category Dynamics Rules

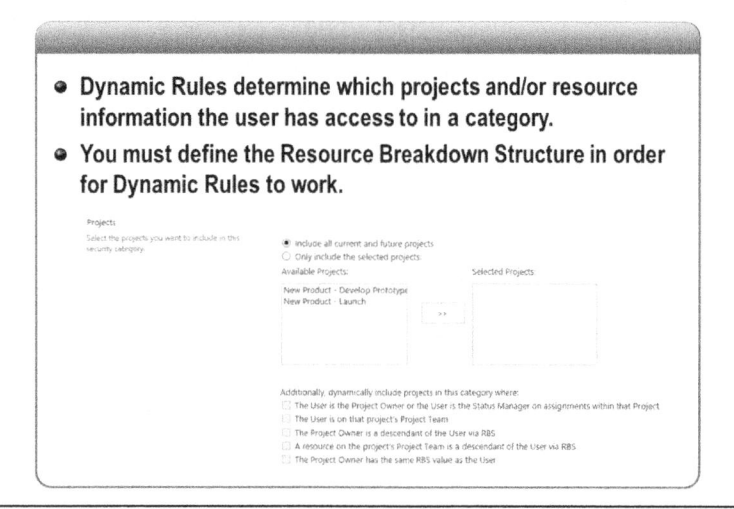

- Dynamic Rules determine which projects and/or resource information the user has access to in a category.
- You must define the Resource Breakdown Structure in order for Dynamic Rules to work.

One area of confusion for some administrators is the Dynamic Rules feature found in Categories. Dynamic Rules determine which projects and/or resource information the user has access to in a category. The key to using dynamic rules is that you must define the RBS. Even if you selected some of the rules, without defining an RBS, these rules will not work and the users in the category may not see or work in the selected Views.

As previously stated, the RBS is an enterprise resource field that is typically used to represent a management hierarchy within an organization. This can be structured by either business unit/department or by geographic locations. Once the RBS is defined, an RBS value needs to be assigned to each resource.

For example, Robert Cain is the Director of Technology for TRIMAGNA Corporation and has an RBS value of TRIMAGNA.TECH. Jan Kalis is the Development Manager that reports to Robert and has an RBS value of TRIMAGNA.TECH.DEV. Tiberiu Covaci is a Software Developer that reports to Robert and has an RBS value of TRIMAGNA.TECH.DEV.DEVELOPER.

If the Project Online administrator selected the Resource Dynamic Rule, "They are direct descendants of the User via RBS", because Robert's RBS value is higher it enables him to view only Jan's resource information.

If the Project Online administrator selected the Resource Dynamic Rule, "They are descendants of the User via RBS", then Robert would see Jan and Tiberiu's resource information.

Key Notables

- You must have an RBS defined and RBS values assigned to resources in order to use dynamic rules.
- With dynamic rules, you do not need to explicitly add projects to a category. It is automatically done for you based on the RBS values for the resources.

Project Dynamic Rules

Rule Name	Description	Default Categories
The User is the Project Owner or the User is the Status Manager on assignments within that Project	Gives users permissions on any project they own	My Personal Projects My Projects
The User is on that project's Project Team	Gives users permissions on any project where they are on the project team. Users do not need to have assignments on the project	My Projects My Tasks
The Project Owner is a descendant of the User via RBS	Gives users permissions on any project that is managed by resources under them in the RBS hierarchy	None
A resource on the project's Project Team is a descendant of the User via RBS	Allows a user to view any project where a resource is under them in the RBS and the resource is on the project team	My Projects
The Project Owner has the same RBS value as the User	The user can view projects managed by people that have the same RBS value	None

Resource Dynamic Rules

Rule Name	Description	Default Categories
The User is the resource	Gives users permissions to view information about themselves (such as assignments)	My Personal Projects My Projects My Tasks
They are members of a Project Team on a project owned by the User	Gives users permissions to view information for all resources in projects they own	My Projects
They are descendants of the User via RBS	Gives users permissions to view information for all resources under them in the RBS	My Resources
They are direct descendants of the User via RBS	Gives users permissions to view information about resources that are directly under them in the RBS	My Direct Reports
They have the same RBS value as the User	Gives users permissions to view information about resources that have the same RBS value	None

Security Templates

- **Default templates that have pre-defined set of global and category permissions**
  - Administrator
  - Portfolio Manager
  - Portfolio Viewer
  - Project Manager
  - Proposal Reviewer
  - Resource Manager
  - Team Lead
  - Team Member

Security Templates are a quick way of defining and applying global and category permissions to users, groups and categories. The benefit to using security templates is that you can standardize your organization's permissions according to the user or group's role. Another benefit to using security templates is that it can eliminate possible errors caused by selecting individual permissions for each user or group.

During the installation of Project Online, the following seven default security templates are created. Each default template has pre-defined set of global and category permissions.

- Administrator
- Portfolio Manager
- Portfolio Viewer
- Project Manager
- Proposal Reviewer
- Resource Manager
- Team Lead
- Team Member

NOTE: *You can apply Allow and Deny settings for both global and category permissions in security templates.*

Practice: Creating Security Templates

> **In this practice, you will:**
> - Create a Security Template for the PMO
> - Modify the Project Manager's Default Security Template

In this practice, you will create a security template that will be used on a security group and category to be created for the Project Management Office (PMO).

Afterwards, you will you will create a copy of the default Project Manager's security template and then modify the default Project Manager's security template.

Exercise 1: Create a Security Template for the PMO

In this exercise, you will create the security template to be used for the PMO group.

1. On the **PWA Settings** page, under the **Security** section, click **Manage Security Templates**.
2. On the **Manage Templates** page, in the menu bar, click **New Template**.
3. On the **Add or Edit Template** page, complete the following settings:

Setting	Perform the following:
Template Name	Type **PMO Manager**
Description	Type **Custom Project Online permission template for PMO Managers**
Copy Template	Select **Portfolio Manager**

4. In the **Message from webpage** dialog box, click **OK**.

   This will pre-populate the category and global permissions with the Portfolio Manager template.

5. Modify the following permissions and click **Save**:

Module 2: Managing Project Online Security

Category Permissions

Project section

Setting	Perform the following:
Create Deliverable and Legacy Item Links	Select **Allow** check box
Edit Project Summary Fields	Select **Allow** check box

Resource section

Setting	Perform the following:
Adjust Timesheet	Select **Allow** check box
Approve Timesheets	Select **Allow** check box
Manage Resource Delegates	Select **Allow** check box

Global Permissions

Admin section

Setting	Perform the following:
Clean Up Project Server Database	Select **Allow** check box
Manage Active Directory Settings	Select **Allow** check box
Manage Gantt Chart and Grouping Formats	Select **Allow** check box
Manage Security	Select **Allow** check box
Manage Users and Groups	Select **Allow** check box

Time and Task Management section

Setting	Perform the following:
Manage Rules	Select **Allow** check box
Manage Time Reporting and Financial Periods	Select **Allow** check box
Manage Time Tracking	Select **Allow** check box
View Project Timesheet Line Approvals	Select **Allow** check box
View Resource Timesheet	Select **Allow** check box

Exercise 2: Modify the Project Manager's Default Security Template

In this exercise, you will create a copy of the default Project Manager's security template and then modify the default Project Manager's security template.

1. On the **Manage Templates** page, in the menu bar, click **New Template**.
2. On the **Add or Edit Template** page, complete the following settings:

Setting	Perform the following:
Template Name	Type **COPY of Project Manager**
Description	Type **Backup Copy of original Project Online permission template for Project Managers**
Copy Template	Select **Project Manager**

3. In the **Message from webpage** dialog box, click **OK**.

   This will pre-populate the category and global permissions with the Portfolio Manager template.

4. On the **Add or Edit Template** page, click **Save**.
5. On the **Manage Templates** page, under **Template Name**, click **Project Manager**.
6. On the **Add or Edit Template** page, under the **Name** section, in the **Description** box, type **MODIFIED Custom Project Online permission template for Project Managers**.
7. In the **Category Permissions** section, change the following settings and click **Save**:

Category Permissions	
Project section	
Setting	Perform the following:
Delete Project	Clear check box under **Allow**
Edit Project Summary Fields	Clear check box under **Allow**

NOTE: *Remember that not assigning Allow or Deny setting, implicitly does not give permission.*

Lesson 4: Creating Project Online Security Principals

- Creating Project Online Users
- Understanding Project Online Groups
- Creating Project Online Groups
- Understanding Project Online Categories
- Creating Project Online Categories

Project Online users, groups and categories are security principals, in that they can be granted permissions to view or use security objects. As a security principal, it maintains a list of security attributes that need to be defined in order to be created.

In this lesson, you will learn how to create Project Online users. You will learn how Project Online groups work and how to create them. You will also learn how Project Online categories work and how to create them.

Objectives

After completing this lesson, you will be able to:

- Create Project Online Users
- Understand how Project Online Groups work
- Create Project Online Groups
- Understand how Project Online Categories work
- Create Project Online Categories

Creating Project Online Users

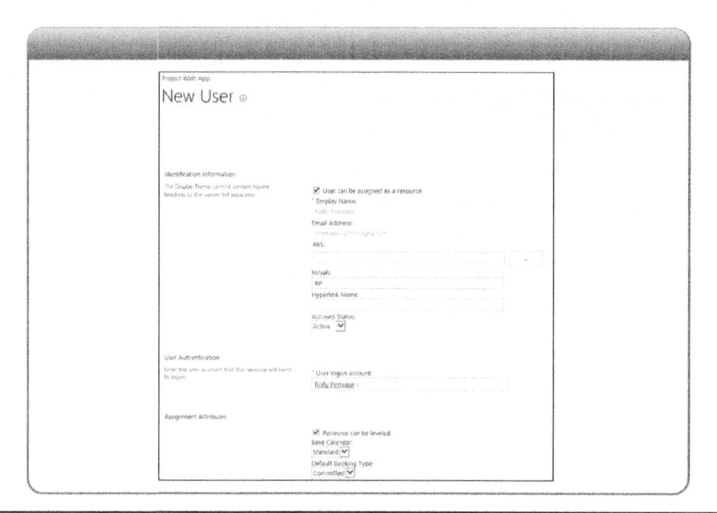

A Project Online user is any individual who is authenticated and has been authorized to access PWA or Project Online Professional. Project Online users are authenticated by either Office 365 or Azure Active Directory methods. Each user must be assigned permission to view or access the data in a particular area of PWA, Project Online, or Project Online Professional. You can directly assign user permissions, or you can assign the users as members to groups and then assign permissions to the group.

How to Create Project Online Users using PWA

1. On the **PWA Home** page, from the **Settings** menu, click **PWA Settings**.
2. On the **PWA Settings** page, in the **Security** section, click **Manage Users**.
3. On the **Manage Users** page, in the menu bar, click **New User**.
4. On the **New User** page, in the **Identification Information** section, do the following, as needed:
   - Select "**User can be assigned as a resource**" check box, if you want the user to be added as an **Enterprise Resource**. This setting is selected by default.
     o If you select this option, when you save the new user, by default you will also create the user as an active work resource in the enterprise resource pool.
   - In the **Display Name** text box, the field is greyed out. This field will be populated after the User logon account is entered. This is a required field.
   - In the **E-mail address** text box, the field is greyed out. This field will be populated after the User logon account is entered and if they also have an Office 365 mailbox This is an optional field.
   - In the **RBS** box, click on the ellipse button (...) to display the RBS Code list to select a value for the user. This is an optional field.
   - In the **Initials** box, type the initials for the user. This is an optional field.
   - If the user maintains a team web site or a SharePoint Server personal MySite, type the hyperlink name field to be displayed and the URL address field.

5. In the **User Authentication** section, in the **User logon account** box, type the Office 365 user name.

6. In the **Assignment Attributes** section, enter the following information about the user's ability to be assigned to projects as a resource:

    - To require approval to use this resources in projects, select the **Resource requires approval for all project assignments** check box. This setting is cleared by default.

    - To exclude this user from the leveling process, clear the **Resource can be leveled** check box. This setting is selected by default.

    > *What is Leveling?*
    > *Leveling is the process of resolving resource conflicts or over-allocations by delaying or splitting certain tasks. If a resource manager levels a resource, its selected assignments are distributed and rescheduled.*

    - In the **Base Calendar** list, select the base calendar that you want to assign to the user. **Standard Calendar** is selected by default.

    - In the **Default Booking Type** list, select the user's booking type as either committed or proposed. **Committed** is selected by default.

    > *What is the difference between a Committed Resource and a Proposed Resource?*
    > *A committed resource is a resource that is formally allocated to any task assignments they have within a project; a proposed resource is a resource with a pending resource allocation to a task assignment that has not yet been authorized.*

    - In the **Timesheet Manager** box, type or search for the manager's name that will be responsible for approving the user's timesheet. The user's display name is entered by default.

    - In the **Default Assignment Owner** box, type or search for the assignment owner's name. The user's display name is entered by default.

    > *What is an Assignment Owner?*
    > *An assignment owner is the person responsible for entering progress information in PWA. This person can be different than the person initially assigned to the task.*

    - To specify the user's availability, type the dates in the **Earliest Available** box and the **Latest Available** box.

    - To set the user's **Standard** and **Overtime** rates, type the value in the respective boxes.

    - To set the **Current Max. Units**, type the percentage value.

    > *What is Max. Units?*
    > *Max. Units is the percentage of time that the resource is available for assignments.*

    - To set the **Cost/Use**, type the value.

    > *What is Cost/Use?*
    > *The per-use cost of the resource if applicable. For work resources, a per-use cost accrues every time the resource is used. For material resources, a per-use cost is accrued only one time.*

7. In the **Departments** section, select a resource's department, if configured.

8. In the **Security Groups** section, in the **Available Groups** list, select the group(s) that the user will become a member and click >. The **Team Members** group is added by default.

9. In the **Security Categories** section, in the **Available Categories** list, select the category that the user will access and click >. Afterwards, when you select the corresponding category in the **Selected Categories** list, the permissions for the selected category are displayed.

   IMPORTANT: *You will need to select each category and grant the necessary permissions. You can apply category permissions manually or by using a security template.*

10. To set category permissions using a template, select the template from the **Set permissions with Template** list and click **Apply**.

    BEST PRACTICE: *You should set category permissions to the group and have the user be a member of the group. Configure category permissions for individual users on rare occasions.*

11. In the **Global Permissions** section, select the global permissions for the user. You can apply global permissions manually or by using a security template. To set global permissions using a template, select the template from the **Set permissions with Template** list and click **Apply**.

    BEST PRACTICE: *You should set global permissions to the group and have the user be a member of the group. Configure global permissions for individual users on rare occasions for special requirements that are not covered by the global permissions that are assigned to a group.*

12. In the **Group Fields** section, if your organization created codes for grouping and costing purposes, type the codes in the corresponding **Group** box, **Code** box, **Cost Center** box, or **Cost Type** box.

13. In the **Team Details** section, if you want the user to be a member of an existing team, in the **Team Name** box, click the ellipse (...) and select the defined team name. If the user is also the **Team Lead**, select **Team Assignment Pool** check box.

14. In the **System Identification** section, in the **External ID** box, you can type additional user identification information. For example, the ID could be the company employee number assigned from the human resources department.

Understanding Project Online Groups

- **Benefits to adding users to groups**
  - Users automatically inherit the permissions of any group that they are added to.
  - Significantly reduce the amount of time spent managing individual user permissions.
  - Users can belong to multiple groups.
  - Groups can be configured to synchronize with security groups in Office 365.

Project Online groups are collections of Project Online users who have similar roles, same security requirements, require similar information or functionality needs. The key thing to remember about creating a group is that it can be based on anything, as long as users in the group will share something in common and will require the same permissions.

Key benefits to using groups:

- Users automatically inherit the permissions of any group that they are added to
- Significantly reduce the amount of time spent managing individual user permissions
- Users can belong to multiple groups
- Groups can be configured to synchronize with security groups in Office 365

Default Groups

There are seven groups created by default when Project Online is installed. Each group is preconfigured with the category and global permissions defined for its role.

Administrators Group – Users added in this group are granted all permissions to Project Online. Assigned categories to this group: My Organization.

Executives Group – Users added in this group are primarily granted permissions to view project and resource information in the Project Center, the Resource Center and Data Analysis. Assigned categories to this group: My Organization.

Portfolio Managers Group – Users added in this group are primarily granted the ability to view and edit all projects and resources in the organization. Assigned categories to this group: My Organization.

Project Managers Group – Users added in this group are primarily granted the ability to create and publish projects. Assigned categories to this group: My Organization and My Projects.

Resource Managers Group – Users added in this group are primarily granted the ability to view and edit resource information, as well as view project and assignment information.

Assigned categories to this group: My Direct Reports, My Organization, My Projects and My Resources.

Team Leads Group – Users added in this group are primarily granted the ability to create new proposals and activity plans, manage resource notifications, edit status report requests, and self-assigned tasks. Assigned categories to this group: My Projects.

Team Members Group – All new users created by using PWA are added to this group by default. They are primarily granted the ability to view their projects and edit their assignments in the timesheet view. They are given access to timesheets, status reports and to do list features. Assigned categories to this group: My Tasks.

Creating Project Online Groups

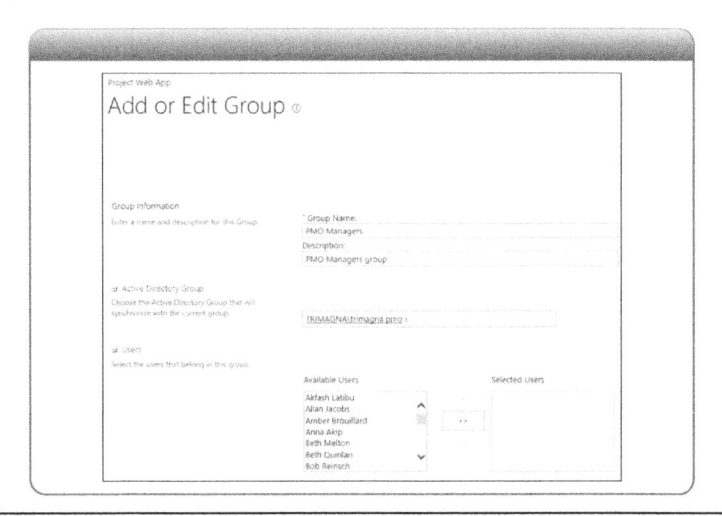

Project Online groups can be granted permissions to view or use security objects. Any users that are added to the group will automatically inherit the group permissions. As a security principal, it maintains a list of security attributes that need to be defined.

How to Create a Project Online Group

1. On the **PWA Home** page, from the **Settings** menu, click **PWA Settings**.
2. On the **PWA Settings page**, in the **Security** section, click **Manage Groups**.
3. On the **Manage Groups** page, click **New Group**.
4. On the **Add or Edit Group** page, in the **Group Information** section:
   - In the **Group Name** box, type the name of the new group
   - In the **Description** box, type a brief description of the groups
   - If you are synchronizing this Project Online group with an Office 365 group, type the name of the security or distribution group
5. In the **Users** section, in the **Available Users** list, select the users who will belong to the group and click >.
6. In the **Categories** section, in the **Available Categories** list, select the categories that the group can access and click >.

   For each category in the Selected Category list, select the permissions that you want the group members to have when they access the category. You can set permissions manually or you can apply a predefined security template.

7. In the **Global Permissions** section, select the global permissions for the group. You can apply global permissions manually or by applying a predefined security template. To set global permissions using a template, select the template and click **Apply**.

Understanding Project Online Categories

- **Categories determine what users can see and do**
- **Default Categories**
  - My Direct Reports
  - My Organization
  - My Projects
  - My Resources
  - My Tasks

Categories are a unique security entity in Project Online. As previously discussed, they are a security object because they can be secured by granting category permissions to users or groups. But they are also a security principal because they have their own set of permissions (category permissions) that can be granted permissions on security objects.

There are five categories created by default when Project Online is installed. Each category has default groups assigned and is preconfigured with the category permissions defined for its role.

My Direct Reports – This category ensures that its members can view all the people they manage directly. That is the people who are on the next level down in the organizational breakdown structure as defined in the resource breakdown structure (RBS) field.

My Organization – This category ensures that its members can view any project, task, resource and assignment published to the server.

My Projects – This category ensures that all projects where a project manager or particular user works on can see all views that are defined for those projects.

My Resources – This category ensures that its members can view all projects that are worked on by the resources that report to them in the organizational breakdown structure, as defined in the RBS field.

My Tasks – This category ensures that its members can see their assignments as well as all projects that they are assigned to the project center.

Creating Project Online Categories

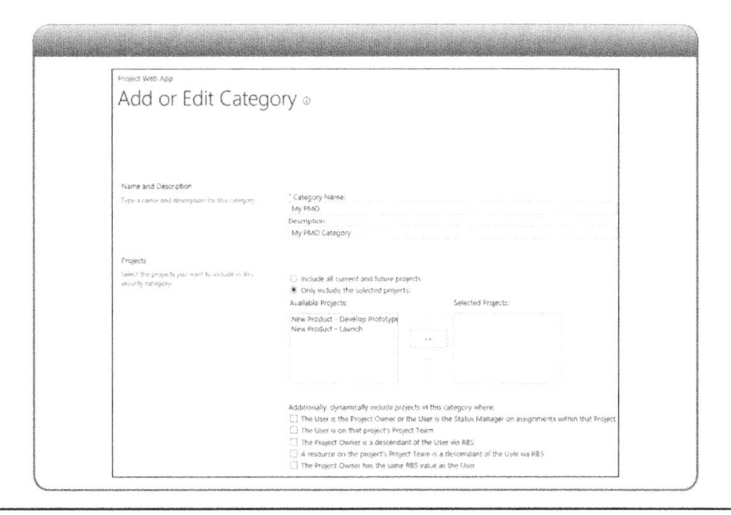

In Project Server permission mode, categories are the collections of projects, resources, and views to which users and groups in PWA are granted access. Categories define which collections of specific data (projects, resources, and views) that these users and groups have access to. Categories also allow the administrator to filter data using security rules, like RBS, that can help organize and display data in specific ways.

You can add projects and resources to categories manually by choosing them from lists, or you can use dynamic filters to automatically add them to categories. Any user associated with a category can be granted permission to the projects and resources in that category.

How to Create a Project Online Category

1. On the **PWA Home** page, from the **Settings** menu, click **PWA Settings**.
2. On the **PWA Settings page**, in the **Security** section, click **Manage Categories**.
3. On the **Manage Groups** page, click **New Category**.
4. On the **Add or Edit Category** page, in the **Group Information** section:
    - In the **Category Name** box, type the name of the new category
    - In the **Description** box, type a brief description of the category
5. In the **Projects** section, specify the projects that users and groups in this category can view. There are two ways that users and groups who have access to this category can view projects:
    - **Include all current and future projects** – Select this option if you want all users and groups within this category to see all projects.
    - **Only include the selected projects** – Select this option if you want all users and groups within this category to see only specific projects. Select the project and click >.
    - If you want category project permissions that are based on the organization's RBS enterprise field, select the check box(es) that best represents the relationship a user must have with a project in order to view the project within the category.

6. In the **Resources** section, specify the resources that users and groups in this category can view. There are three ways that users and groups who have access to this category can view resource information:

    - **Include all current and future resources** – Select this option if you want all users and groups within this category to see all resources.

    - **Only include the selected resources** – Select this option if you want all users and groups within this category to see only specific resources. Select the resource and click >.

    - If you want category resource permissions that are based on the organization's RBS enterprise field, select the check box(es) that best represents the relationship a user must have with users in order to view the resources within the category.

7. In the **Views** section, select the views the users and groups in this category can see by selecting the view name check box.

8. In the **Permissions** section, select the users or groups that will be assigned to the category and click >. Once the user or group has been added, select the category permissions that you want the selected user or group to have when they access the category.

    You can set category permissions manually or you can apply a predefined security template. To set category permissions using a template, select the template and click **Apply**.

Practice: Working with Project Online Security Principals

> **In this practice, you will:**
> - Create a Project Online User Account
> - Create a Project Online Group
> - Apply a Security Template to a Project Online Group
> - Create a Project Online Category
> - Apply a Security Template to a Project Online Category

In this practice, you will create an PWA user account and then create a PWA group and configure the Active Directory Synchronization feature. You will then create a new Project Online category for the PMO Managers and then modify a category and apply custom security templates.

Exercise 1: Create a Project Online User Account

In this exercise, you will create a new Project Online user account.

1. On the **Manage Templates** page, click **Settings** and click **PWA Settings**.
2. On the **PWA Settings page**, in the **Security** section, click **Manage Users**.
3. On the **Manage Users** page, in the menu bar, click **New User**.
4. On the **New User** page, complete the configuration using the following settings and click **Save**:

User Authentication section	
Setting	**Perform the following:**
User logon account	Type **Brian Rieck**

Identification Information section	
Setting	**Perform the following:**
Display Name:	Should display **Brian Rieck**
Initials:	Type **BR**

5. On the **Manage Users** page, verify that the user account is listed.

Exercise 2: Create a Project Online Group

In this exercise, you will create a Project Online group and add a Project Online user to this group.

1. Click **Settings** and click **PWA Settings**.
2. On the **PWA Settings page**, under **Security** section, click **Manage Groups**.
3. On the **Manage Groups** page, in the menu bar, click **New Group**.
4. On the **Add or Edit Group** page, complete the configuration using the following settings and click **Save**:

   Group Information section

Setting	Perform the following:
Group Name:	Type **PMO Managers**
Description:	Type **CUSTOM PMO Managers group**

   Active Directory Group section

Setting	Perform the following:
Active Directory Group	Type **PMO Managers Grp**

   Global Permissions section

Setting	Perform the following:
Set permission with Template	Select **PMO Manager** and click **Apply**

5. On the **Manage Groups** page, click **Portfolio Managers**.
6. On the **Add or Edit Group** page, in the **Active Directory Group** section, type **Business Analysts Grp** and click **Save**.
7. Repeat steps 5-6 for the following Project Online group names:

Project Online Group	Office 365 Security Group to Synchronize
Portfolio Viewers	Executives Grp
Project Managers	Project Managers Grp
Resource Managers	Resource Managers Grp
Team Leads	Team Leads Grp
Team Members	Team Members Grp

8. On the **Manage Groups** page, on the menu bar, click **Active Directory Group Sync Options**.
9. On the **Sync Project Web App security groups with Active Directory** page, select **Enable Scheduled Synchronization** check box and click **Save and Synchronize Now**.
10. Wait for 10 seconds and press the **F5** key to refresh the web browser.
11. On the **Manage Groups** page, in the **Last Sync** column, the status should display **Succeeded at <date and time>**.

Module 2: Managing Project Online Security

12. Click on a group to verify the names of Project Online user accounts have been added to the groups. When finished reviewing, click **Cancel** to return to the **Manage Groups** page.

Exercise 3: Apply a Security Template to a Project Online Group

In this exercise, you will apply the custom security templates previously created to the associated default Project Online groups.

1. On the **Manage Groups** page, click **PMO Managers**.
2. On the **Add or Edit Group** page, under the **Global Permissions** section, at the bottom of the list, on the **Set permissions with Template** list, select **PMO Manager**, click **Apply** and click **Save**.
3. On the **Manage Groups** page, click **Project Managers**.
4. On the **Add or Edit Group** page, under the **Global Permissions** section, at the bottom of the list, on the **Set permissions with Template** list, select **Project Manager**, click **Apply** and click **Save**.

Exercise 4: Create a Project Online Category

In this exercise, you will create a new Project Online Category.

1. On the **Manage Groups** page, click **Settings** and click **PWA Settings**.
2. On the **PWA Settings page**, under **Security** section, click **Manage Categories**.
3. On the **Manage Categories** page, in the menu bar, click **New Category**.
4. On the **Add or Edit Category** page, complete the configuration using the following settings and click **Save**:

Name and Description section	
Setting	Perform the following:
Category Name:	Type **My PMO**
Description:	Type **CUSTOM My PMO category**

Projects section	
Setting	Perform the following:
Select the projects you want to include in this category	Select **Include all current and future projects**

Resources section	
Setting	Perform the following:

Setting	Perform the following:
Select the resources you want to include in this category	Select **Include all current and future resources**

Views section

Setting	Perform the following:
Select the views you want to add to the category	Select **All** check boxes

Permissions section

Setting	Perform the following:
Available Users and Groups	Select **PMO Managers** and click >
Permissions for PMO Managers (Group)	In the **Set permissions with Template** list, select **PMO Manager**, click **Apply**

Exercise 5: Apply a Security Template to a Project Online Category

In this exercise, you will apply a custom security template to the My Projects category.

1. On the **Manage Categories** page, click **My Projects**.

2. On the **Add or Edit Category** page, expand the **Permissions** section, on the **Users and Groups with Permissions** list, select ***Project Managers**,

3. In the **Permissions for Project Managers (Group)** list, in the **Set permissions with Template** list, select **Project Manager** and click **Apply**.

   NOTE: Notice that the checkboxes under **Allow** *for* **Delete Project** *and* **Edit Project Summary Fields** *permissions are cleared.*

4. Click **Save**.

Summary

> **In this module, you learned how (to):**
> - Project Online security works
> - Work in SharePoint Permission Mode
> - Work in Project Permission Mode
> - Create Project Online Security Entities:
>   - Project Online Users
>   - Project Online Groups
>   - Project Online Categories

In this module, you learned how Project Online security works for properly securing access to sensitive project and resource data. Project Online security is divided into two types of security entities: security principals and security objects.

You learned about SharePoint permission mode and Project Online permission mode and how to configure both. You also learned about Project Online security principals and how to create Project Online users, groups and categories.

Objectives

After completing this module, you learned how (to):

- Project Online security works
- Work in SharePoint Permission Mode
- Work in Project Permission Mode
- Create Project Online Security Entities:
  - Project Online Users
  - Project Online Groups
  - Project Online Categories

This page is intentionally left blank

Module 3:
Working with Microsoft Project Clients

Contents

Module Overview 1
Lesson 1: Overview of Project Clients 2
    Overview of Project Clients – On-Premises 3
    Overview of Project Clients – Cloud-Based 4
    Installing Microsoft Project Professional 2016 5
    Installing Microsoft Project Online Professional 6
    Practice: Installing Project Online Professional 7
Lesson 2: Configuring Project Clients 9
    Difference between Authentication and Authorization 10
    Creating a Project Web App Profile 11
    Signing in to Project Online 12
    Managing Local Project Cache 13
    Practice: Configuring Project Online Professional 15
Lesson 3: Using Project Web App 17
    Configuring User Access to Project Web App 18
    Supported 3rd Party Web Browsers 19
    Signing in to Project Web App 20
    Impersonating Users 21
    Signing Out of Project Web App 23
    Practice: Using Project Web App 24
Summary 25

EXCLUSIVELY PUBLISHED BY

PMO Logistics
679 Roberta Avenue
Winnipeg, Manitoba, Canada R2K 0K9

Copyright © 2017 by Roland Perreaux

All rights reserved. No part of the contents of this document may be reproduced or transmitted in any form or by any means without written permission of the publisher.

Information in this document, including URL and other Internet Web site references, is subject to change without notice. Unless otherwise noted, the example companies, organizations, products, domain names, e-mail addresses, logos, people, places, and events depicted herein are fictitious, and no association with any real company, organization, product, domain name, e-mail address, logo, person, place, or event is intended or should be inferred. Complying with all applicable copyright laws is the responsibility of the user. Without limiting the rights under copyright, no part of this document may be reproduced, stored in or introduced into a retrieval system, or transmitted in any form or by any means (electronic, mechanical, photocopying, recording, or otherwise), or for any purpose, without the express written permission of PMO Logistics Inc.

The names of manufacturers, products, or URLs are provided for informational purposes only and PMO Logistics makes no representations and warranties, either expressed, implied, or statutory, regarding these manufacturers or the use of the products with any Microsoft technologies. The inclusion of a manufacturer or product does not imply endorsement of Microsoft of the manufacturer or product. Links are provided to third party sites. Such sites are not under the control of PMO Logistics and PMO Logistics is not responsible for the contents of any linked site or any link contained in a linked site, or any changes or updates to such sites. PMO Logistics is not responsible for webcasting or any other form of transmission received from any linked site. PMO Logistics is providing these links to you only as a convenience, and the inclusion of any link does not imply endorsement of PMO Logistics of the site or the products contained therein.

PMO Logistics may have patents, patent applications, trademarks, copyrights, or other intellectual property rights covering subject matter in this document. Except as expressly provided in any written license agreement from PMO Logistics, the furnishing of this document does not give you any license to these patents, trademarks, copyrights, or other intellectual property.

PMO Logistics, Professional Training Series, Upgrading Skills Series, TriMagna Corporation and TriMagna Corporation logo are either registered trademarks or trademarks of PMO Logistics Inc. in Canada, the United States and/or other countries.

Microsoft, Active Directory, Internet Explorer, Outlook, Project Server, SharePoint, SQL Server, Visual Studio, Windows and Windows Server are either registered trademarks or trademarks of Microsoft Corporation in the United States and/or other countries.

All other trademarks are property of their respective owners.

Author:	Rolly Perreaux, PMP, MCSE, MCT
Publisher:	PMO Logistics
Developmental Editor:	Heather Perreaux
Cover Graphic Design:	Andrea Ardiles
Technical Testing:	Underground Studioworks

Post-Publication:
Errata List Contributors:

Module Overview

- Overview of Project Clients
- Configuring Project Clients
- Using Project Web App

In this module, you will learn how to install and configure Project Online Professional to work with Project Online. You will also learn how to use Project Web App.

Objectives

After completing this module, you will be able to:

- Install both editions of Microsoft Project clients
  - On-Premises
  - Cloud Based
- Configure Microsoft Project Clients
- Use Project Web App

Lesson 1: Overview of Project Clients

- Overview of Microsoft Project 2016 Clients – On-Premises
- Overview of Microsoft Project Online Clients – Cloud-Based
- Installing Microsoft Project Professional 2016
- Installing Microsoft Project Online Professional

In this lesson, you will learn how to install and configure Project Online Professional and Project Professional 2016 and the difference between authentication and authorization. You will also learn how to create a Project Web App Profile, log in using the new profile and manage the local project cache.

Objectives

After completing this lesson, you will be able to:

- Understand the difference between Microsoft Project Online Professional and Project Professional 2016
- Install Microsoft Project Online Professional and Project Professional 2016

Overview of Project Clients – On-Premises

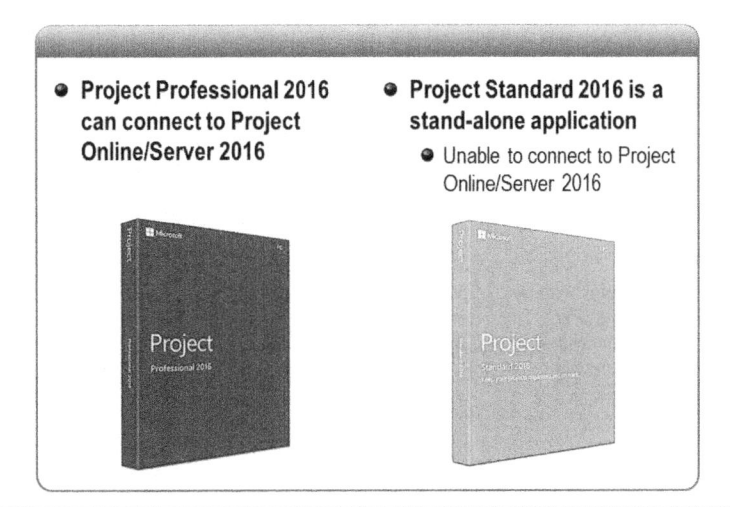

- **Project Professional 2016 can connect to Project Online/Server 2016**
- **Project Standard 2016 is a stand-alone application**
  - Unable to connect to Project Online/Server 2016

Microsoft offers two editions of Project 2016:

- Microsoft Project Professional 2016 Edition
- Microsoft Project Standard 2016 Edition

Project Professional 2016 has all the same tools and features found in Project Standard 2016. However, the difference between the two editions is that Project Standard 2016 is a stand-alone desktop application. Stand-alone means that the software is not connected to external servers.

Project Professional 2016 can be used as a desktop application and can also connect to Project Server 2016 and Project Online.

Overview of Project Clients – Cloud-Based

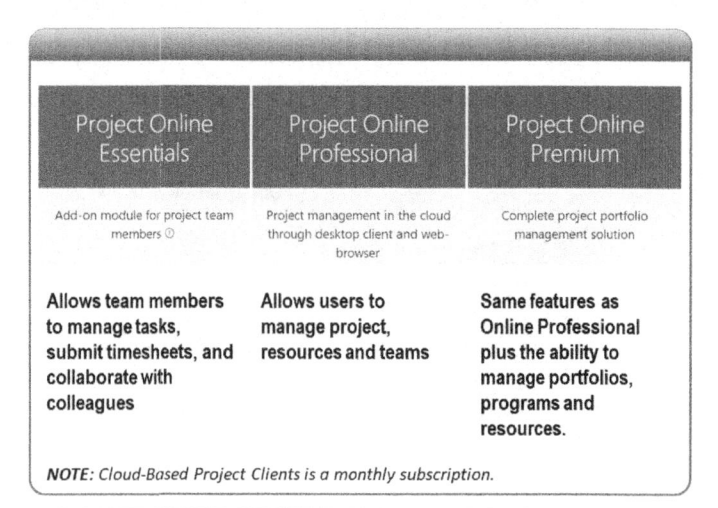

Project Online Essentials	Project Online Professional	Project Online Premium
Add-on module for project team members	Project management in the cloud through desktop client and web-browser	Complete project portfolio management solution
Allows team members to manage tasks, submit timesheets, and collaborate with colleagues	Allows users to manage project, resources and teams	Same features as Online Professional plus the ability to manage portfolios, programs and resources.

NOTE: *Cloud-Based Project Clients is a monthly subscription.*

Microsoft offers three cloud-based client solutions for Project Online:

Project Online Essentials

Specifically for project team members to collaborate with colleagues in the cloud, via web browser or mobile device. Team members are also allowed to manage their tasks and submit their timesheets.

NOTE: *There is no desktop client for Project Online Essentials*

Project Online Professional

Specifically for project managers and resource managers to manage their projects, resources and teams in the cloud through desktop client and/or web browser.

Project Online Premium

Offers the same features as Project Online Professional, but also provides the ability to manage portfolios, programs and resources. This solution is a complete project and portfolio management solution.

Installing Microsoft Project Professional 2016

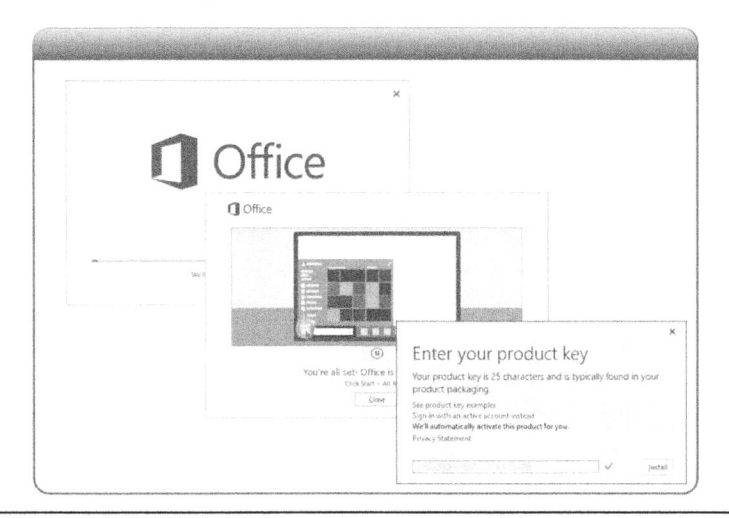

Installing Microsoft Project Professional 2016 is slightly different than in previous versions, but it is also a much easier process.

1. Simply insert your **Project Professional 2016** disc into the CD/DVD ROM and the setup program should automatically start installing the software.

2. There are no installation options to pick. Essentially the software automatically installs.

3. When you receive the **You're all set! Office is installed now**, click **Close**.

4. From the **Start** menu, launch **Project 2016**.

5. At the **Enter your product key** dialog box, enter your Project 2016 product key and click **Install**.

6. Project Professional 2016 opens and the installation is complete.

7. Click **File → Account** to view the Product Information.

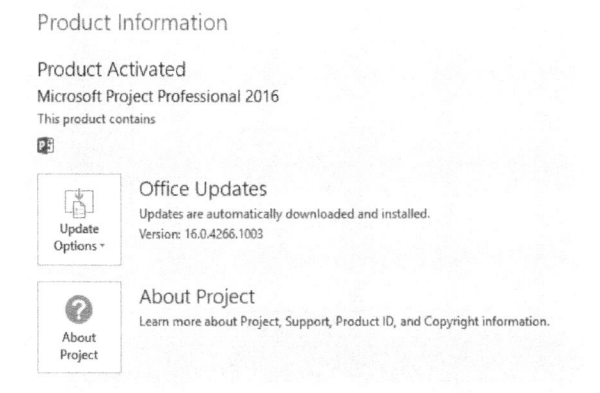

Installing Microsoft Project Online Professional

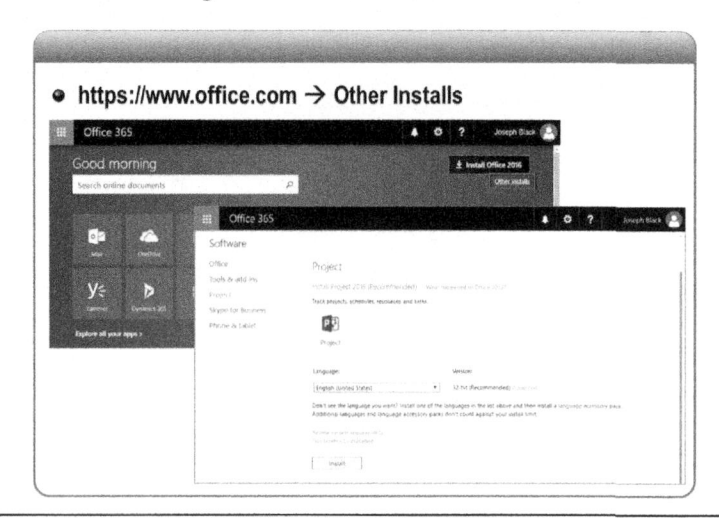

Installing Microsoft Project Online Professional is done via the Office website and is a very simple process.

1. Launch a web browser and in the URL box, type https://www.office.com.
2. On the **Office 365** page, in the upper right corner, click on **Other Installs**.
3. On the **Software** page, on the side menu, click **Project**, then click **Install**.
4. In the **Just a few more steps…** window, a setup file is automatically downloaded to your computer. Just follow the instructions as shown below to install Microsoft Project Online Professional.

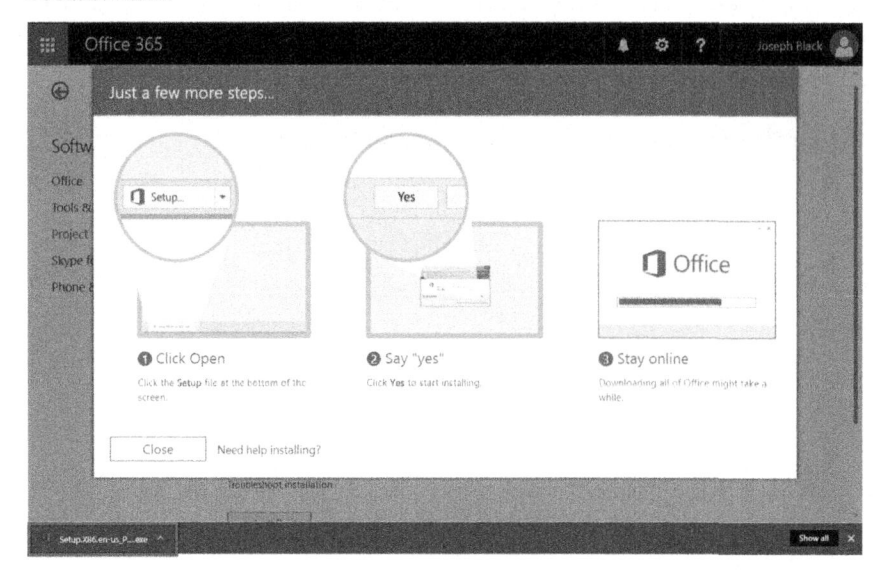

Practice: Installing Project Online Professional

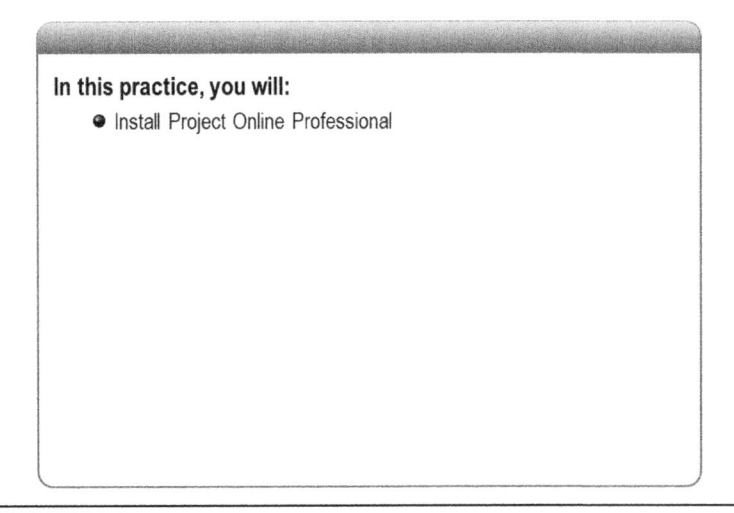

In this practice, you will:
- Install Project Online Professional

Exercise 1: Install Project Online Professional

In this exercise, you will install Project Online Professional from the Office.com website.

1. From Internet Explorer in the URL box, type https://www.office.com.

   NOTE: Sign in as the Project Online Administrator

2. On the **Office 365** page, in the upper right corner, click on **Other Installs**.

   *NOTE: You may have a different Office 365 page, in that you click the arrow next to **Install Office apps** and then click **Other Install Options**.*

3. On the **Software** page, on the side menu, click **Project**, then click **Install**.

4. In the **Just a few more steps…** window, a setup file is automatically downloaded to your computer. Just follow the instructions to install Microsoft Project Online Professional.

5. At the bottom of the browser, click the **Setup** file and click **Yes**.

   NOTE: If you have any other Microsoft Office applications running, the installation process will close all opened apps.

   Microsoft Project starts downloading.

6. When you see the **You're all set! Office is installed now** window, click **Close**.

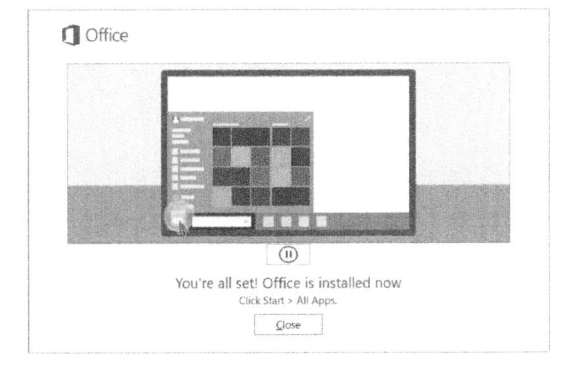

7. From the **Start** menu, click **Project 2016**.
8. On the **Office is almost ready** dialog box, click **Accept and start Project**.

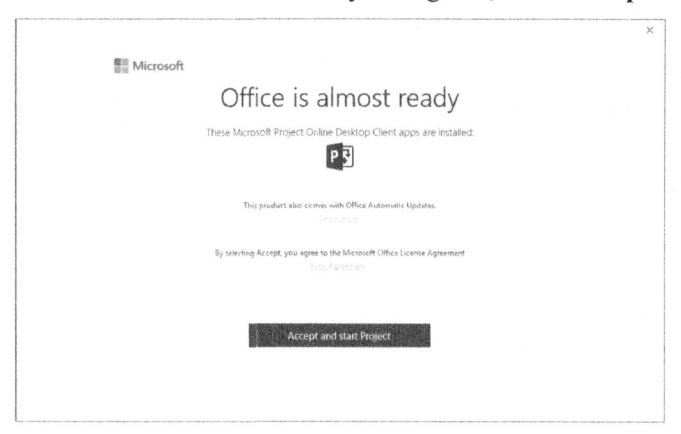

9. On the **Project – Recent** page, click **Blank Project**.
10. From the **File** menu, click **Account.**
11. Review the **Product Information** section, when finished, click the back button.

Lesson 2: Configuring Project Clients

- Difference between Authentication and Authorization
- Creating a Project Web App Profile
- Signing in to Project Online
- Managing Local Project Cache

In this lesson, you will learn how to configure Project Online Professional and Project Professional 2016 and the difference between authentication and authorization. You will also learn how to create a Project Web App Profile, log in using the new profile and manage the local project cache.

Objectives

After completing this lesson, you will be able to:

- Understand the difference between authentication and authorization
- Create a Project Web App Profile
- Log on to Project Online using Project Online Professional
- Manage Local Project Cache

Difference between Authentication and Authorization

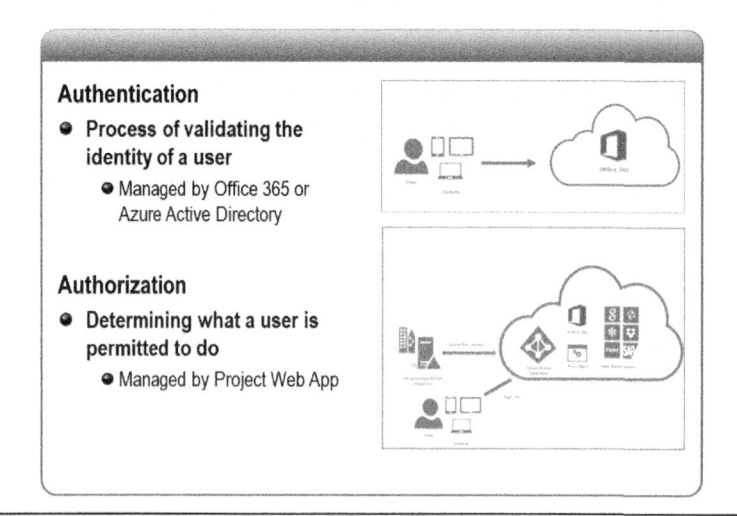

Authentication
- Process of validating the identity of a user
  - Managed by Office 365 or Azure Active Directory

Authorization
- Determining what a user is permitted to do
  - Managed by Project Web App

Authentication is the process of validating the identity of a user, whereby authorization is the Authentication is the process of validating the identity of a user, whereby authorization is the process of determining what a user is permitted to do. Think of it this way, if Project Online was a building, authentication is your key to the front door, and authorization is what you are allowed to do once inside using the same key.

After the user identity is validated, the authorization process determines which sites, content, and other features the user can access.

By default, Project Online uses Office 365 user authentication. So, when you initially connect to Project Online using Project Professional or Project Web App, you will be using an Office 365 user account.

NOTE: If you currently have Active Directory on-premises, you can implement Azure AD Connect to allow your users to have Single Sign-On (SSO) capability to access Microsoft Office 365 (including Project Online)

After Office 365 performs authentication of the user, the security features in Project Online perform the authorization process. Authorization is discussed in greater detail in the next module Managing Project Online Security.

What is Azure AD Connect?

Azure AD Connect is a tool for synchronizing directory data between Active Directory on-premises and Azure AD in the cloud. It allows you to provide a common identity for your users for Office 365, Azure, and SaaS applications integrated with Azure AD.

NOTE: Windows Azure Active Directory Sync (DirSync) or Azure AD Sync are deprecated tools and are not supported.

More information on Azure AD Connect can be found at:
https://docs.microsoft.com/en-us/azure/active-directory/connect/active-directory-aadconnect

Creating a Project Web App Profile

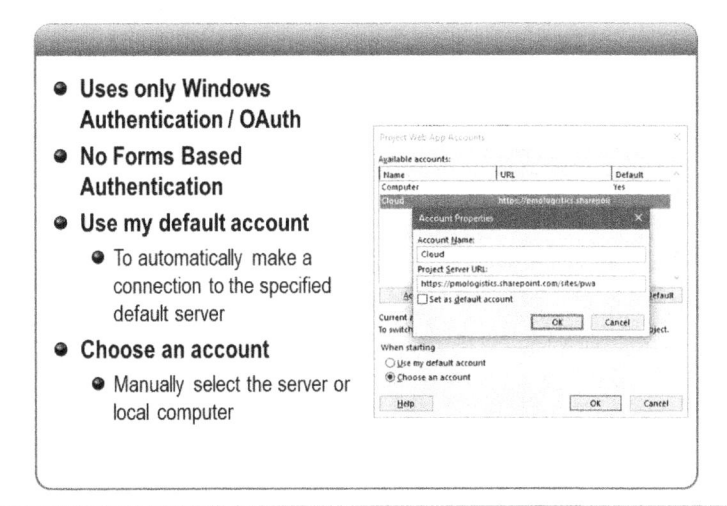

- **Uses only Windows Authentication / OAuth**
- **No Forms Based Authentication**
- **Use my default account**
  - To automatically make a connection to the specified default server
- **Choose an account**
  - Manually select the server or local computer

Authentication profiles are the mechanism for Project Online Professional users to be authenticated to Project Online via PWA.

1. Launch **Project Online Professional**, from the **File** menu, click **Info** and click **Manage Accounts**.

2. In the **Project Web App Accounts** dialog box, click **Add**.

3. In the **Account Properties** dialog box, complete the following:

    a. In the **Account Name** box, type a unique name for this account.

    b. In the **Project Server URL** box, type the URL of the PWA.

4. In the **Project Web App Accounts** dialog box, select the preference on how you connect to Project Online in the **When starting** section:

    - Select **Use my default account**, if you want Project to automatically detect and make a connection to the specified default server. If a default server is not specified, then Project will open offline with no server connections.

    - Select **Choose an account**, if you prefer to select a server to connect to each time you open Project. Use this option if you have multiple servers you frequently access, or if you sometimes prefer to use Project offline.

5. Click **OK** and restart **Project Online Professional**.

Signing in to Project Online

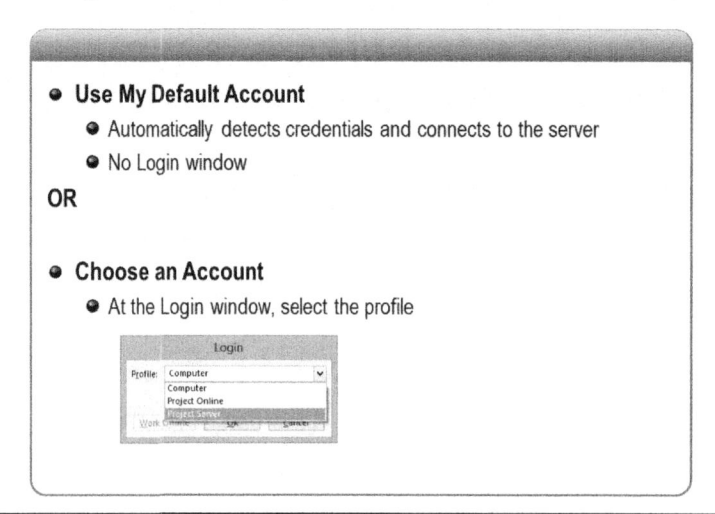

- **Use My Default Account**
  - Automatically detects credentials and connects to the server
  - No Login window

OR

- **Choose an Account**
  - At the Login window, select the profile

Once the Project Web App Profile is created, you simply need to start Project Online Professional and select the profile that was created in the **Login** dialog box to connect to Project Online.

If you have multiple authentication profiles, you will be able to select the profile of your choice using the drop-down list.

If you receive a "Project Server Unavailable" message when you try to connect to Project Online, verify with the Project Online Administrator that you are in a Project group that has permission to log in from Project Online Professional.

NOTE: By default, only members of the Administrators and Project Managers groups have permission to log on from Project Online Professional.

Managing Local Project Cache

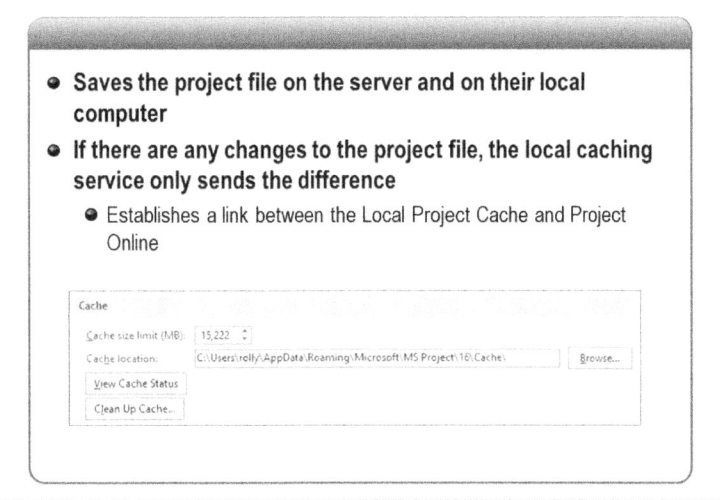

> - **Saves the project file on the server and on their local computer**
> - **If there are any changes to the project file, the local caching service only sends the difference**
>   - Establishes a link between the Local Project Cache and Project Online
>
> Cache
> Cache size limit (MB): 15,222
> Cache location: C:\Users\rolly\AppData\Roaming\Microsoft\MS Project\16\Cache\   Browse...
> View Cache Status
> Clean Up Cache...

Local Project Cache helps project managers to work more efficiently by synchronizing a local version of their project with the version that is stored on Project Online.

When project managers use Project Professional to create a project, they will save the project on Project Online. This not only saves the file on the server, but also saves a local version on their computer to the Local Project Cache (aka Active Cache).

Anytime a Project Manager saves their project to Project Online, it will save the project locally in the Local Project Cache and then use the computer's local caching service to send only the difference between the original local save and the second local save to Project Online. The benefits are:

- The local caching service establishes the communications link between the Local Project Cache and Project Online.
- Allows Project Online Professional and the project manager to perform other work instead of waiting for the save process to complete. This can be a huge time saving.
- The save process is performed by using an asynchronous process instead of synchronous process.

An asynchronous process is the data communications which can be transmitted intermittently instead of a steady stream. For example, when speaking with someone on the telephone the conversation would be considered as an asynchronous process because both parties can talk whenever they like. If you were speaking with someone on a radio the conversation would be considered as a synchronous process, because in order to talk, you would have to wait for the other person on the radio to finish talking. This is why radio operators say "Over" when they are finished speaking to cue the other radio operator to begin speaking.

How to Adjust the Cache Settings

1. From the **File** menu, click **Options**.
2. In the **Project Options** dialog box, in the side menu, click **Save**.
3. In the **Cache** section, in the **Cache size limit (MB)** box, type the maximum amount of memory (in megabytes) that you want the cache to occupy on your hard drive.

4. In the **Cache location** box, type the path to the location that you want to use for the cache or click **Browse** and navigate to the new location.

How to Remove Projects from the Cache

Removing projects from the cache can be helpful in reducing the number of projects you see in the Open dialog box, or in reducing the amount of memory used when approaching the cache size limit.

1. From the **File** menu, click **Options**.
2. In the **Project Options** dialog box, click **Save**.
3. In the **Cache** section, click **Clean Up Cache**.
4. In the **Remove projects from cache** section, in the **Project Filter** list, you can select either Projects not checked out to you or Projects checked out to you to display the corresponding list of projects.
5. Select the project that you want to remove from the cache and click **Remove From Cache**.

How to View the Cache Status

You can view your recent cache activities, such as checking a project in or saving changes to Project Online, in the Active Cache Status dialog box. If you receive an error while synchronizing a project with Project Online, you can view error messages on the Active Cache Status dialog box.

1. From the **File** menu, click **Options**.
2. In the **Project Options** dialog box, in the navigation pane, click **Save**.
3. In the **Cache** section, click **View Cache Status**.
4. On the **Status** tab, review your recent cache activities, including the date and time when each activity occurred, and the status of the activity.
5. On the **Errors** tab, review any errors you may have received while synchronizing your projects with Project Online.

Practice: Configuring Project Online Professional

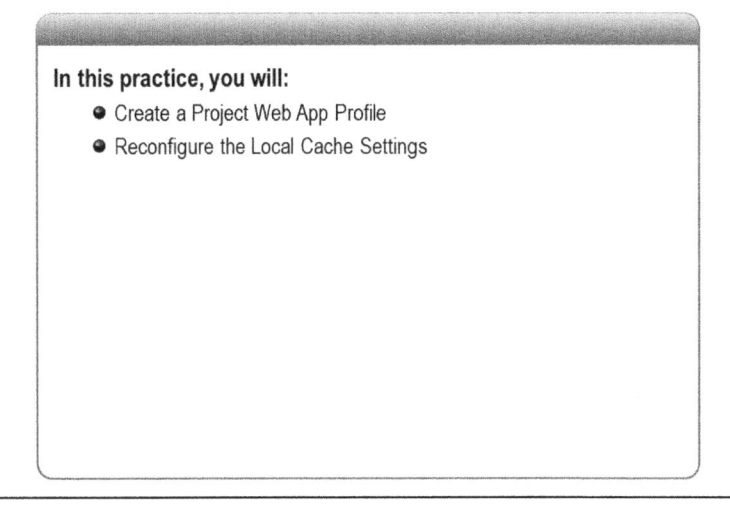

> **In this practice, you will:**
> - Create a Project Web App Profile
> - Reconfigure the Local Cache Settings

In this practice, you will create a Project Web App profile and then configure the local cache settings in Project Online Professional.

Exercise 1: Create Project Web App Profile

In this exercise, you will create Project Web App profile to connect to Project Online.

1. From the **File** menu, select **Info** and click **Manage Accounts**.
2. In the **Project Web App Accounts** dialog box, click **Add**.
3. In the **Account Properties** dialog box, and complete the following settings and click **OK**.

Setting	Perform the following:
Account Name	Type **Project Online**
Project Server URL	Type the URL to your Project Online tenant
Set as default account	Select check box

4. In the **Microsoft Project** dialog box, review the message and click **Yes**.
5. In the **Project Web App Accounts** dialog box, in the **When starting** section, select **Choose an account** and click **OK**.
6. Close and restart **Project Online Professional**.
7. In the **Login** dialog box, accept **Project Online** and click **OK**.

   As this is the first time connecting to the PWA site, this may take a minute.

8. On the **Project – Recent** page, click on **Blank Project**.
9. At the bottom left corner, if you hover your mouse cursor over the globe, the message should display **Connected to Project Online**, as shown below:

Exercise 2: Reconfigure Local Cache Settings

In this exercise, you will reconfigure the Local Cache Settings in Project Online Professional.

1. From the **File** menu, click **Options**.
2. In the **Project Options** dialog box, in the navigation pane, click **Save**.
3. In the **Cache** section, in the **Cache location**, click **Browse**.
4. In the **Modify Location** window, navigate to **Local Disk (C:)**, create a new folder called **MS Project Cache**, double click the new folder and click **OK**.
5. The new cache location should be **C:\MS Project Cache**, click **OK**.
6. In the **Microsoft Project** dialog box, click **OK**.
7. Close **Project Online Professional**.

Lesson 3: Using Project Web App

- Configuring User Access to Project Web App
- Supported 3rd Party Web Browsers
- Signing in to Project Web App
- Impersonating Users
- Signing Out to Project Web App

In this lesson, you will learn how to use Project Online using a web browser. You will also learn how to sign in as a different user and know the list of supported 3rd party web browsers for PWA.

Objectives

After completing this lesson, you will be able to:

- Configuring User Access to Project Web App
- Know the supported 3rd party web browsers
- Sign in to Project Web App
- Impersonate Users
- Sign out to Project Web App

Configuring User Access to Project Web App

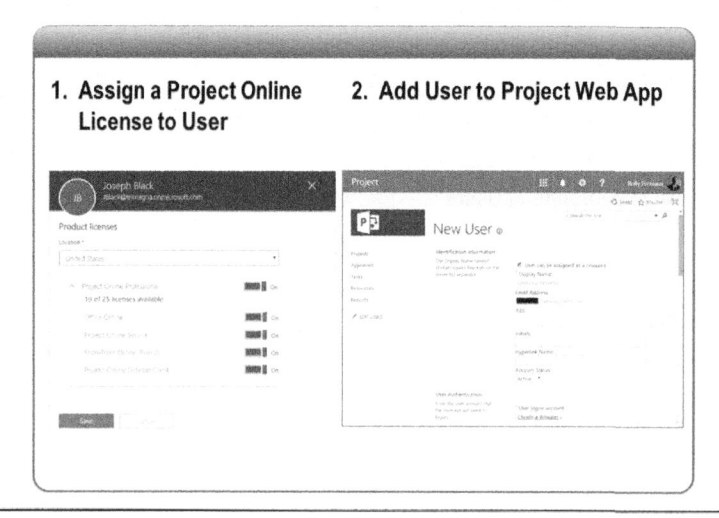

In order to sign in to the Project Web App site, the Project Online Administrator must complete two tasks:

User is Assigned a Project Online License

The user must be assigned a Project Online license in order to be authenticated. If the user is not assigned a Project Online license, they will receive the following error message:

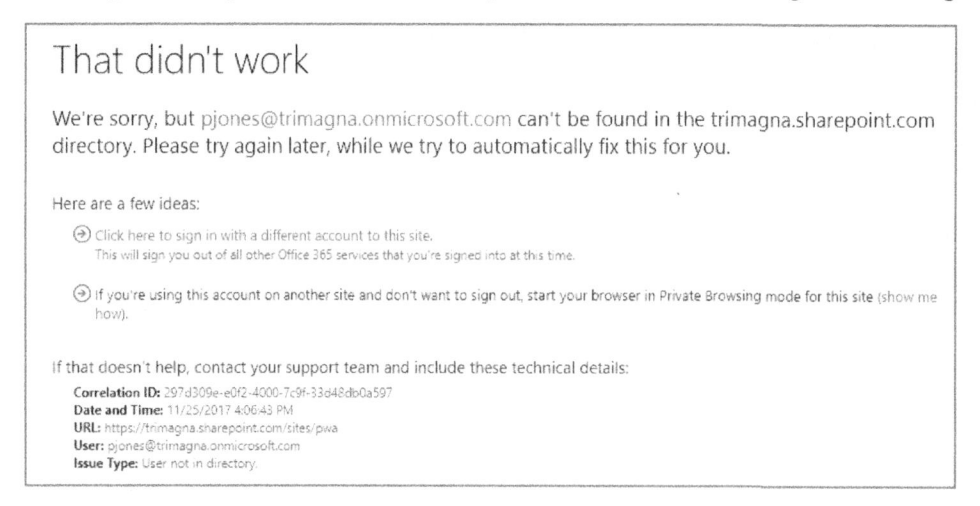

> **That didn't work**
>
> We're sorry, but pjones@trimagna.onmicrosoft.com can't be found in the trimagna.sharepoint.com directory. Please try again later, while we try to automatically fix this for you.
>
> Here are a few ideas:
>
> → Click here to sign in with a different account to this site.
> This will sign you out of all other Office 365 services that you're signed into at this time.
>
> → If you're using this account on another site and don't want to sign out, start your browser in Private Browsing mode for this site (show me how).
>
> If that doesn't help, contact your support team and include these technical details:
> Correlation ID: 297d309e-e0f2-4000-7c9f-33d48db0a597
> Date and Time: 11/25/2017 4:06:43 PM
> URL: https://trimagna.sharepoint.com/sites/pwa
> User: pjones@trimagna.onmicrosoft.com
> Issue Type: User not in directory.

PWA User Account is created in Project Online

The user must be assigned a Project Online license in order to be authorized. If the user is not created in Project Online, they will receive the following error message:

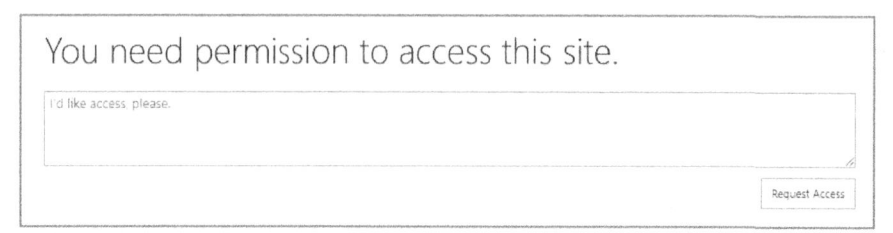

> **You need permission to access this site.**
>
> I'd like access, please.
>
> Request Access

Supported 3rd Party Web Browsers

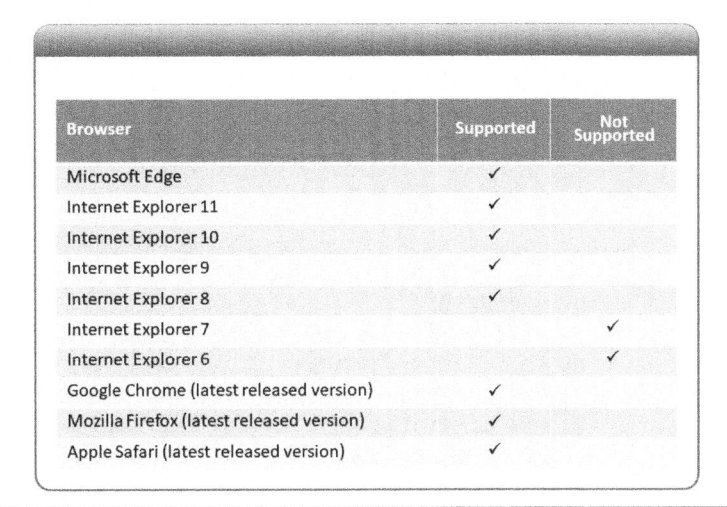

Browser	Supported	Not Supported
Microsoft Edge	✓	
Internet Explorer 11	✓	
Internet Explorer 10	✓	
Internet Explorer 9	✓	
Internet Explorer 8	✓	
Internet Explorer 7		✓
Internet Explorer 6		✓
Google Chrome (latest released version)	✓	
Mozilla Firefox (latest released version)	✓	
Apple Safari (latest released version)	✓	

Project Online supports the following web browsers:

- Microsoft Edge
- Internet Explorer
- Google Chrome
- Mozilla Firefox
- Apple Safari

However, certain web browsers could cause some SharePoint Online functionality to be downgraded, limited, or available only through alternate steps:

Browser	Supported	Not Supported
Microsoft Edge	✓	
Internet Explorer 11	✓	
Internet Explorer 10	✓	
Internet Explorer 9	✓	
Internet Explorer 8	✓	
Internet Explorer 7		✓
Internet Explorer 6		✓
Google Chrome (latest released version)	✓	
Mozilla Firefox (latest released version)	✓	
Apple Safari (latest released version)	✓	

Signing in to Project Web App

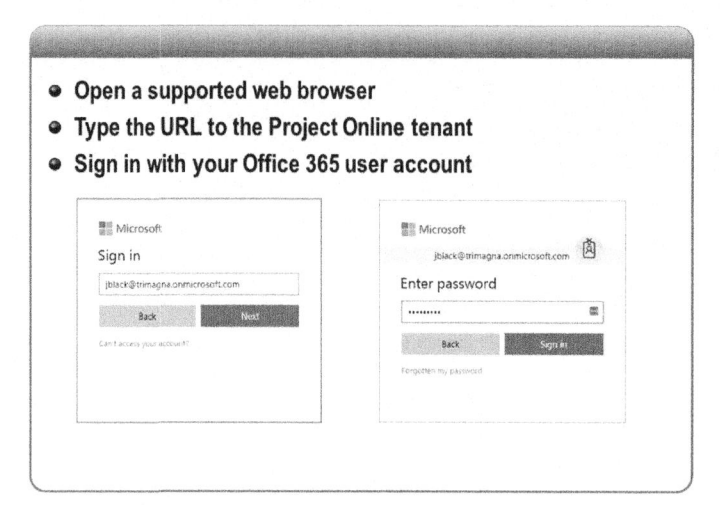

- Open a supported web browser
- Type the URL to the Project Online tenant
- Sign in with your Office 365 user account

Once the Project Online Administrator has completed the two user access tasks, then signing in is a simple process:

1. Open a supported web browser
2. Type the URL to the Project Online tenant
3. Sign in with your Office 365 user account

Impersonating Users

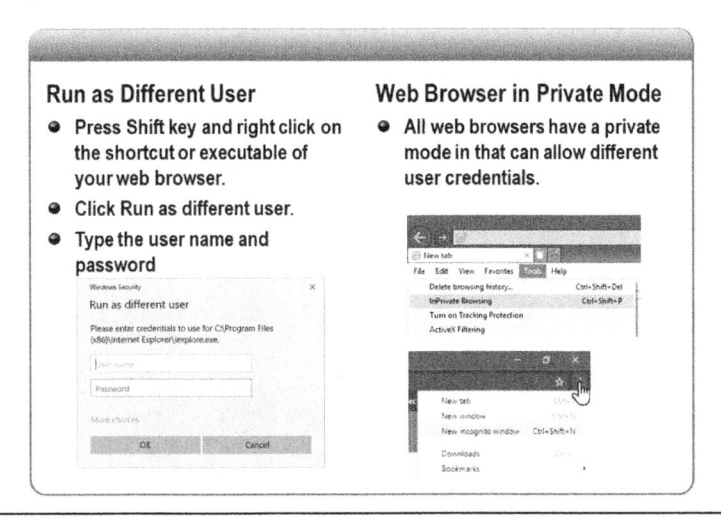

Run as Different User	Web Browser in Private Mode
• Press Shift key and right click on the shortcut or executable of your web browser. • Click Run as different user. • Type the user name and password	• All web browsers have a private mode in that can allow different user credentials.

Impersonating a user is one of many tools that a Project Online Administrator can test user, group and category permissions. You can achieve impersonating a user in two ways:

Web Browser in Private Mode

All supported web browsers allow for some form of a private mode. The key benefit of having your browser in private mode, is that the credentials from a current session do not pass through in private mode. Therefore, you can access the PWA site using different credentials than the one you currently have open.

Internet Explorer

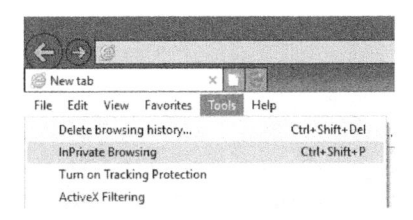

Google Chrome

Firefox

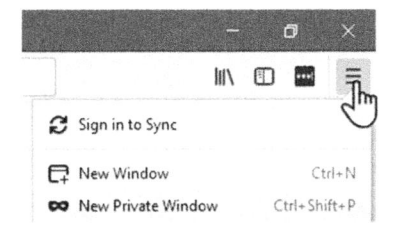

Run as Different User

If your network uses Azure AD Connect you might be able to use the Run as a Different option.

1. From the **Start** page, press the **Shift** key and right click **Internet Explorer**.
2. Then click **Run as different user**, as shown below:

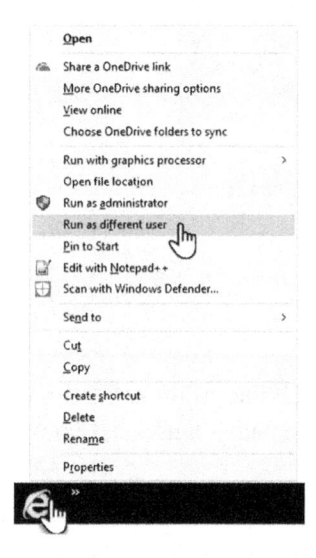

3. In the **Windows Security** dialog box, in the **User name** box, type the user name and the corresponding password and click **OK**.

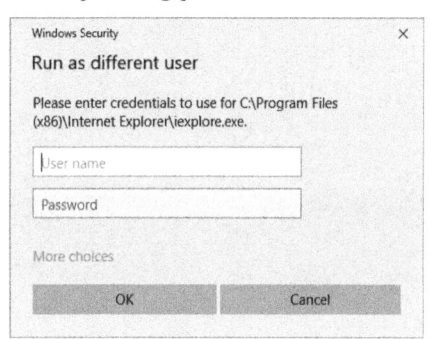

4. In **Internet Explorer**, in the **Address** box, type the URL of the PWA site.

Signing Out of Project Web App

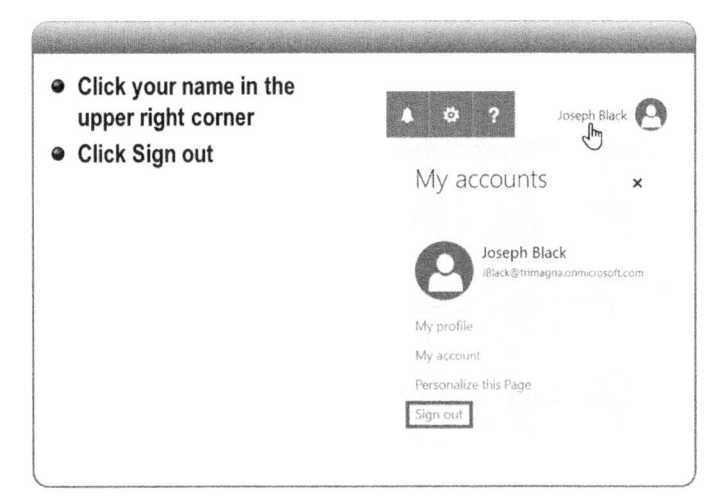

- Click your name in the upper right corner
- Click Sign out

It's always a good practice to sign out when you are finished working in the Project Web App site.

How to Sign Out

1. On the **PWA** page, in the upper right side, click your name
2. Click **Sign Out**

Practice: Using Project Web App

> **In this practice, you will:**
> - Sign in to Project Web App in Private Mode
> - Sign out of Project Web App

In this practice, you will impersonate a Project Online user by signing in to Project Web App in private mode.

Exercise 1: Sign in to Project Web App in Private Mode

In this exercise, you will use the private mode of Internet Explorer and use different user credentials to sign in to PWA.

1. Launch **Internet Explorer**, press **Ctrl + Shift + P** keys to open Internet Explorer in **InPrivate** mode.
2. In the **Address** box, type: **https://<companyname>.sharepoint.com/sites/pwa** and press **Enter**.
3. In the **Sign in** window, type **BQuinlan@<companyname>.onmicrosoft.com** and click **Next**.
4. In the **Enter password** window, type **Pa$$w0rd** and click **Sign in**.
5. Review the PWA home page

Exercise 2: Sign out of Project Web App

In this exercise, you will sign out of Project Web App.

1. On the **PWA Home** page, in the upper right side, click **Beth Quinlan**.
2. Click **Sign Out.**
3. Close **Internet Explorer**.

Summary

> **In this module, you learned how to:**
> - Install both editions of Microsoft Project clients
>   - On-Premises
>   - Cloud Based
> - Configure Microsoft Project Clients
> - Use Project Web App

In this module, you learned how to install and configure Project Online Professional to work with Project Online. You also learned how to use Project Web App in a web browser.

Objectives

After completing this module, you learned how to:

- Install both editions of Microsoft Project clients
  - On-Premises
  - Cloud Based
- Configure Microsoft Project Clients
- Use Project Web App

This page is intentionally left blank

Module 4: Configuring Project Online

Contents

Module Overview 1
Lesson 1: Configuring Time & Task Management Settings 2
    Configuring Fiscal Periods 3
    Configuring Time Reporting Periods 5
    Creating Line Classifications 6
    Configuring Timesheet Settings and Defaults 7
    Configuring Administrative Time 10
    Configuring Task Settings and Display... 11
    Manage Timesheets 13
    Timesheet Managers 14
    Practice: Configuring Time & Task Management Settings 15
Lesson 2: Configuring Operational Policies 19
    Configuring Additional Server Settings 20
    Active Directory Enterprise Resource Pool Synchronization 22
    Configuring Project Site Creation 23
    Configuring Connected SharePoint Sites 25
    Practice: Configuring Operational Policies 26
Lesson 3: Importing Resources and Project Plans 29
    Overview of Importing Data 30
    Importing Resources to Enterprise 31
    Importing Projects to Enterprise 32
    Practice: Importing Resources and Projects 34
Summary 38

EXCLUSIVELY PUBLISHED BY

PMO Logistics
679 Roberta Avenue
Winnipeg, Manitoba, Canada R2K 0K9

Copyright © 2017 by Roland Perreaux

All rights reserved. No part of the contents of this document may be reproduced or transmitted in any form or by any means without written permission of the publisher.

Information in this document, including URL and other Internet Web site references, is subject to change without notice. Unless otherwise noted, the example companies, organizations, products, domain names, e-mail addresses, logos, people, places, and events depicted herein are fictitious, and no association with any real company, organization, product, domain name, e-mail address, logo, person, place, or event is intended or should be inferred. Complying with all applicable copyright laws is the responsibility of the user. Without limiting the rights under copyright, no part of this document may be reproduced, stored in or introduced into a retrieval system, or transmitted in any form or by any means (electronic, mechanical, photocopying, recording, or otherwise), or for any purpose, without the express written permission of PMO Logistics Inc.

The names of manufacturers, products, or URLs are provided for informational purposes only and PMO Logistics makes no representations and warranties, either expressed, implied, or statutory, regarding these manufacturers or the use of the products with any Microsoft technologies. The inclusion of a manufacturer or product does not imply endorsement of Microsoft of the manufacturer or product. Links are provided to third party sites. Such sites are not under the control of PMO Logistics and PMO Logistics is not responsible for the contents of any linked site or any link contained in a linked site, or any changes or updates to such sites. PMO Logistics is not responsible for webcasting or any other form of transmission received from any linked site. PMO Logistics is providing these links to you only as a convenience, and the inclusion of any link does not imply endorsement of PMO Logistics of the site or the products contained therein.

PMO Logistics may have patents, patent applications, trademarks, copyrights, or other intellectual property rights covering subject matter in this document. Except as expressly provided in any written license agreement from PMO Logistics, the furnishing of this document does not give you any license to these patents, trademarks, copyrights, or other intellectual property.

PMO Logistics, Professional Training Series, Upgrading Skills Series, TriMagna Corporation and TriMagna Corporation logo are either registered trademarks or trademarks of PMO Logistics Inc. in Canada, the United States and/or other countries.

Microsoft, Active Directory, Internet Explorer, Outlook, Project Server, SharePoint, SQL Server, Visual Studio, Windows and Windows Server are either registered trademarks or trademarks of Microsoft Corporation in the United States and/or other countries.

All other trademarks are property of their respective owners.

Author: Rolly Perreaux, PMP, MCSE, MCT

Publisher: PMO Logistics
Developmental Editor: Heather Perreaux
Cover Graphic Design: Andrea Ardiles
Technical Testing: Underground Studioworks

Post-Publication:
Errata List Contributors:

Module Overview

- Configuring Time and Task Management Settings
- Configuring Operational Policies
- Importing Resources and Project Plans

In this module, you will learn how to configure time and task settings that include timesheets and the corresponding settings. You will also learn how to configure Project Online operational policies and how to import resources and project plans.

Objectives

After completing this module, you will be able to:

- Configure Time & Task Management Settings
- Configure Operational Policies
- Import Resources and Project Plans

Lesson 1: Configuring Time & Task Management Settings

- Configuring Fiscal Periods
- Configuring Time Reporting Periods
- Creating Line Classifications
- Configuring Timesheet Settings and Defaults
- Configuring Administrative Time
- Configuring Task Settings and Display
- Manage Timesheets
- Timesheet Managers

Before team members can report their time and progress on assignments, you must set up timesheets and task updates to conform to your organizations fiscal and project reporting requirements. In this lesson, you will learn how to configure time and task management settings in Project Online.

Objectives

After completing this lesson, you will be able to:

- Configure Fiscal Periods
- Configure Time Reporting Periods
- Create Line Classifications
- Configure Timesheet Settings and Defaults
- Configure Administrative Time
- Configure Task Settings and Display
- Manage Timesheets
- Define Timesheet Managers

Configuring Fiscal Periods

If you use Project Online timesheets to integrate project time reporting with your company accounting system, you want the financial periods in Project Online to match the fiscal periods that your organization uses.

A fiscal year is the 12-month period a company uses for financial reporting, which does not have to match the calendar year. Companies define fiscal years either to close the books at a convenient time or to follow seasonal variations in business. Because calendar months do not end neatly on the same day of the week, companies use different methods to create consistent fiscal periods within a fiscal year. In Project Online, you can define your organization's fiscal year, quarters, and months.

Manage Fiscal Period

Select a year and then click Define. In that way, you can change the fiscal year you use, for example, to switch to the new fiscal year implemented after a merger with another company.

Define Fiscal Period Start Date

Select the start date for the fiscal year and the method for defining the fiscal periods within the year. You pick the start date for the fiscal year, which automatically sets the end date for fiscal year to the date 12 months later.

Set Fiscal Year Creation Model

Select the option that corresponds to the fiscal periods your organization uses. Regardless of which option you choose, every fiscal year represents 52 weeks. The choices represent different methods of dividing a year into consistent fiscal periods:

4, 5, 4 Method – For each 13-week quarter, this option sets the first month to four weeks, the second month to five weeks, and the third month to four weeks.

4, 4, 5 Method – For each 13-week quarter, this option sets the first and second month in a fiscal quarter to four weeks, and the third month of the quarter to five weeks.

5, 4, 4 Method – For each 13-week quarter, this option sets the first month in a fiscal quarter to five weeks, and the second and third months to four weeks.

13 Months – Sets each fiscal period to four weeks for a fiscal year of 52 weeks.

Standard calendar year – Sets the fiscal quarters to end on March 31, June 30, September 30, and December 31.

NOTE: If your fiscal periods do not match any of the available methods, select the one that is the closest. When you click Create and Save, you will have a chance to adjust the end dates for each fiscal month.

Define Period Naming Convention

Choose a prefix, number, and suffix to uniquely identify each fiscal period. For example, in the setting shown below, results in fiscal month names of M1-08, M2-08, and so on. As you fill in the boxes in this section, the sample at the bottom of the section displays the name of the first fiscal period.

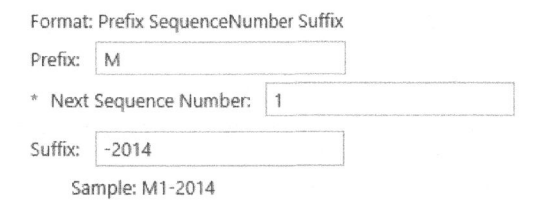

Format: Prefix SequenceNumber Suffix
Prefix: M
* Next Sequence Number: 1
Suffix: -2014
Sample: M1-2014

When you click **Create and Save**, the Fiscal Periods page appears, listing the start and end dates for each fiscal month. To edit these dates to match your organization's fiscal periods, click an End Date, click the calendar icon that appears, and then click the end date for the fiscal period.

Configuring Time Reporting Periods

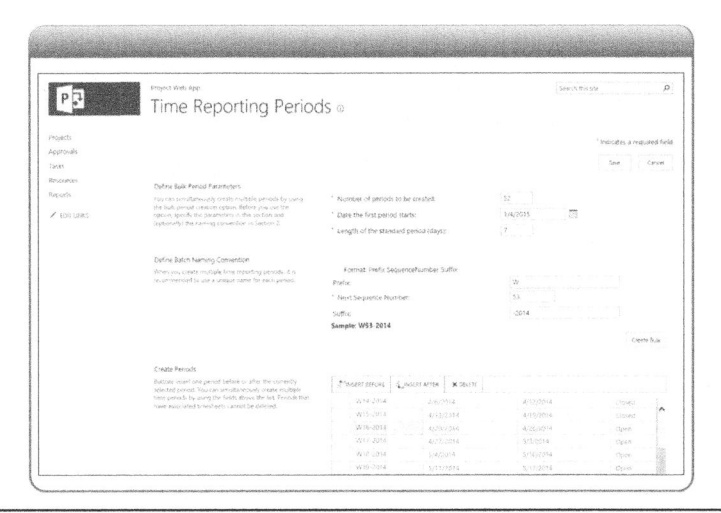

The second aspect of timesheets is the reporting period. Similar to defining fiscal periods, you set up timesheet periods by clicking Time Reporting Periods on the Server Settings page in PWA.

Define Bulk Period Parameters

Specify the time reporting period parameters including the number of time periods you want to create, the starting date for the first period, and the length of each period. By default, the number of periods is set to 52 to create a year's worth of one-week periods.

Define Batch Naming Convention

The boxes in this section define the naming convention for the time periods. For example, setting the prefix to W and the suffix to "-2014" names each weekly time period, W1-2014, W2-2014, and so on.

When you click **Create Bulk**, Project Online creates the time periods with the start and end date.

Creating Line Classifications

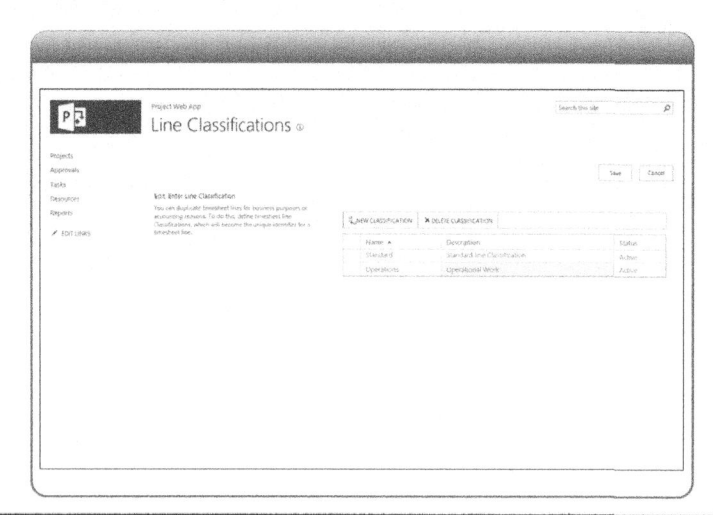

The time that resources work can fall into different billable and non-billable categories. Line classifications represent different types of billable work. Project Online comes with the Standard line classification for project work, but you can create additional line classifications for other types of billable work that you want to track. For example, you may want to track billable work that resources performed on past projects that have transitioned into ongoing maintenance.

How to Create a New Line Classification

1. On the **PWA Settings** page, click **Line Classifications**.

2. On the **Line Classifications** page, click **New Classification** and then type a name and description in the new row that appears in the table.

   *NOTE: When you click **Save**, the new line classification is immediately available.*

3. When a resource adds a line to a timesheet, the new classification appears in the Line Classification box.

Configuring Timesheet Settings and Defaults

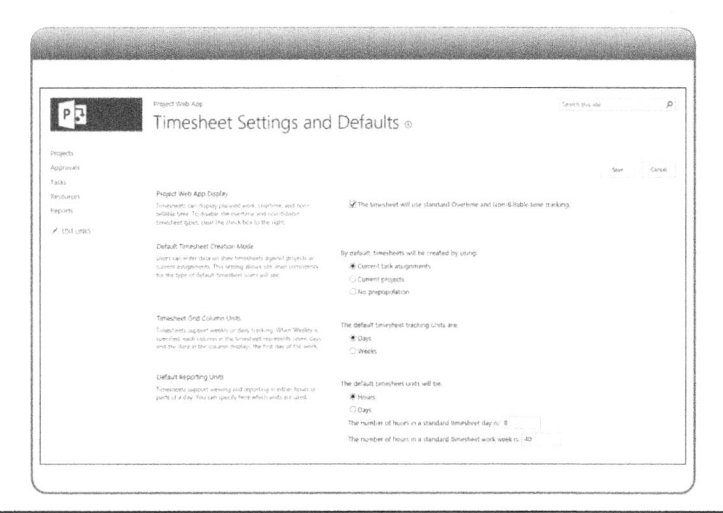

Before resources can submit their time, you must choose settings to display the information you want on timesheets and control how resources can fill timesheets out. You must also define the time periods for timesheets, for example, one week, two weeks, or alternate duration.

To specify timesheet behavior, on the **PWA Settings** page, click **Timesheet Settings and Defaults**.

Project Web App Display

For timesheets displayed in PWA, keep the check box checked to allow resources to record planned project time, overtime, and non-billable time. To disable the overtime and non-billable timesheet types, clear the check box

☑ The timesheet will use standard Overtime and Non-Billable time tracking.

Default Timesheet Creation Mode

Users can enter data on their timesheets against projects or current assignments. This setting allows site-level consistency for the type of default timesheet users will see.

By default, timesheets will be created by using:
- ⦿ Current task assignments
- ○ Current projects
- ○ No prepopulation

Timesheet Grid Column Units

Timesheets support weekly or daily tracking. When Weekly is specified, each column in the timesheet represents seven days, and the date in the column displays the first day of the week.

The default timesheet tracking units are:
- ⦿ Days
- ○ Weeks

Default Reporting Units

The default is Hours, which means that resources report the time they work in hours. Select the Days option to report time in whole or partial days. The default number of hours in a timesheet day is 8 and the default hours in a week is 40. Type different numbers if your organization works shorter or longer days.

> The default timesheet units will be:
> ◉ Hours
> ○ Days
> The number of hours in a standard timesheet day is: 8
> The number of hours in a standard timesheet work week is: 40

Hourly Reporting Limits

Accounting systems, customers, or internal business policies might restrict how time can be entered. If you use team resources, be sure to consider such restrictions when you set these values.

> Time tracking data entry limits:
> Maximum Hours per Timesheet 999
> Minimum Hours per Timesheet 0
> Maximum Hours per Day 999

Timesheet Policies

You can use settings in this section to help your company comply with accounting and/or regulatory policies. You may restrict users from reporting time into the future in their timesheets. You may also enable the functionality to allow unverified timesheet lines. These lines are free form for users to track unstructured time and will not be verified against Project Online projects or tasks. Finally, you may enable Task Status Managers to coordinate or approve/reject timesheet lines on a per line basis. Policy settings only apply to timesheets created after the settings were changed.

> ☑ Allow future time reporting
> ☑ Allow new personal tasks
> ☑ Allow top-level time reporting
> Task Status Manager Approval:
> ○ Enabled
> ☐ Require line approval before timesheet approval
> ◉ Disabled

Auditing

You can use timesheet auditing to record changes saved to timesheets during creation, approval, and later adjustments.

> ☐ Enable Timesheet Auditing   [Purge Log]

Approval Routing

Fixed approval routing will disable the ability to change the next approver during timesheet submission.

☐ Fixed Approval Routing

Single Entry Mode

Select this mode if you want your team members to report project task status in their timesheet.

☐ Single Entry Mode

Configuring Administrative Time

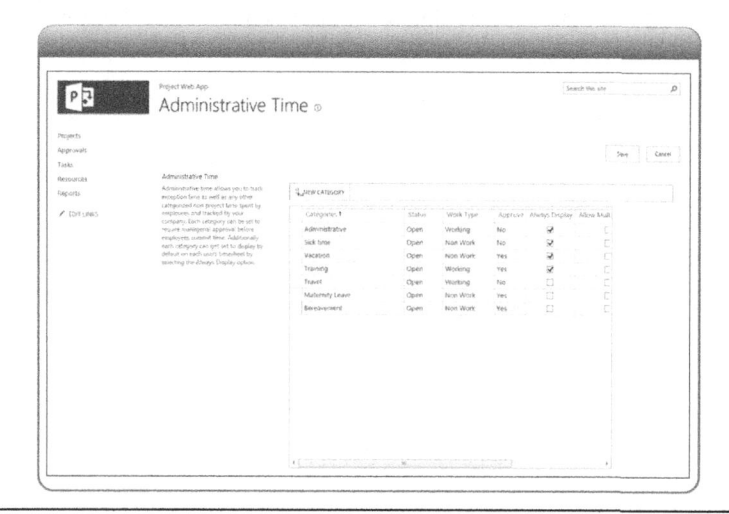

To the dismay of most project managers, resources do not devote every working hour to project work. By setting up categories for administrative time, you can plan for and track time that resources spend on non-project work, whether it is working time, such as training or time spent at meetings, or nonworking time, such as sick time, vacation, or holidays.

Administrative time categories represent different types of non-project work. Project Online comes with the three built-in administrative categories (Administrative, Vacation, and Sick Time), but you can also create additional types.

You can do more than track administrative time that resources have already spent. Resources can request administrative time off, such as vacation time. Then, project or resource managers can evaluate the effect on the project and approve or deny the requests. For approved administrative time, project managers then incorporate the time into resources' calendars and the project schedule.

Categories - Type the name for the new administrative category in the first cell in the row.

Status - By default, a new category is set to Open, which enables resources to record time against the category and is what you want for a category that you have just created. You can change the category to Closed later when you no longer want resources to use the category.

Work Type - By default, a new category is set to **Non Work**, which typically applies to time off that resources receive as part of their benefits, such as holidays, parental leave, or sabbaticals. Choose **Working** for non-project time that resources work, such as training, corporate meetings, or travel.

Approve - New administrative categories are automatically set to **No**, so that resources do not have to ask for approval ahead of time. They can add time for that category to a timesheet for the week in which they took that time. Choose **Yes**, if you want resources to ask for approval, which gives you the chance to approve or deny the request.

Always Display - Check the check box in this cell if you want the category to appear on every timesheet. If you turn the check box off, resources can add the category to a timesheet when they need it.

Configuring Task Settings and Display

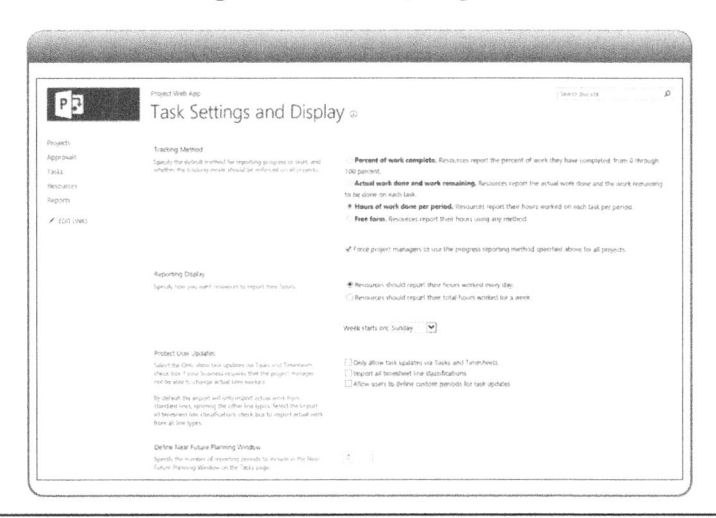

To ensure that resources update tasks with the information you want, you must configure task settings to use the progress update method for your organization. In addition to the update method, you can also specify fields of information that resources can fill in and the circumstances under which resources can modify information.

To configure task settings, open the **PWA Settings** page and click **Task Settings and Display**.

Tracking Method

Each tracking method uses different combinations of fields to measure progress depending on the reporting needs of your project. Each option has its advantages and limitations, and for this reason it is important to analyze your project reporting requirements before choosing one of the following methods:

- **Percent of work complete (*default*)** – Resources report the percent of work they have completed on their tasks. This method is quick and easy, but it does not give project managers the information they need to keep a project on track. A resource could report that a task is 50 percent complete, but you cannot tell whether the resource worked 20 hours and has 20 hours left, or worked 100 hours and has 100 hours to go.

- **Actual work done and work remaining** – Provides more useful information about task progress without requiring detailed time entry. For example, if a resource reports 20 hours worked and 40 hours remaining, you know that the task is one-third complete, but you also know how long it will take to complete.

- **Hours of work done per period** – Resources report the time they work during each time period (days or weeks) of a project, similar to filling in a timesheet. In addition, they can still report how many hours of work remain, so you can schedule the completion of tasks.

- **Free Form** – Resources report their hours using any method.

- **Force project managers to use the progress reporting method specified above for all projects (*default*)** – Requires every project manager to track progress using the same method.

IMPORTANT: *If you are using Single Entry Mode, the Hours of work done per period and Force project managers to use progress reporting method specified above for all projects options are automatically selected and cannot be modified.*

Reporting Display

Resources Should Report Their Hours Worked Every Day option is selected by default. You may obtain more accurate time records with this option, because resources are more likely to remember the work they did that day. To provide more flexibility, select the Resources Should Report Their Total Hours Worked for a Week option.

- ⦿ Resources should report their hours worked every day.
- ◯ Resources should report their total hours worked for a week.

Week starts on: Sunday

Protect User Updates

Select the **Only allow task updates via Tasks and Timesheets** check box if your business requires that the project manager not be able to change actual time worked.

By default the import will only import actual work from standard lines, ignoring the other line types. Select the Import all timesheet line classifications check box to import actual work from all line types.

- ☐ Only allow task updates via Tasks and Timesheets.
- ☐ Import all timesheet line classifications.
- ☑ Allow users to define custom periods for task updates.

Define Near Future Planning Window

Specify the number of reporting periods to include in the Near Future Planning Window on the Tasks page.

Manage Timesheets

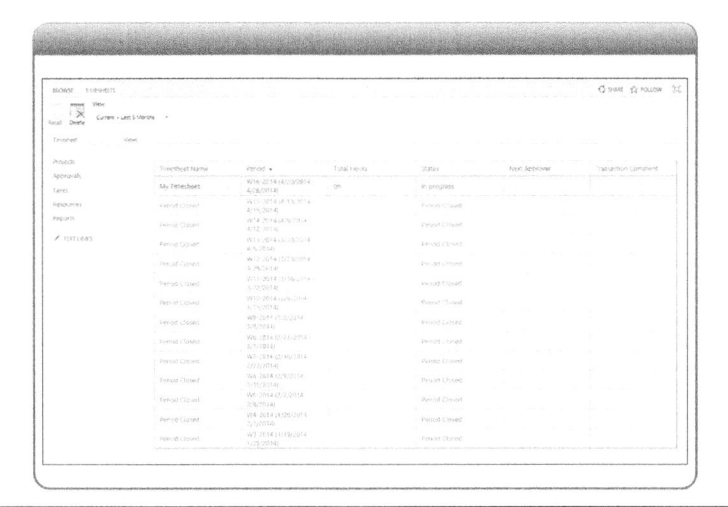

The Manage Timesheets page enables administrators to recall timesheets that have been submitted allowing them to be sent back to the submitter, or delete timesheets that are no longer needed.

To Recall or Delete a Timesheet

1. On the **PWA Settings** page, in the **Time and Task Management** section, click **Manage Timesheets**.

2. Click the row for the timesheet that you want to recall or delete, and then click the corresponding button.

Timesheet Managers

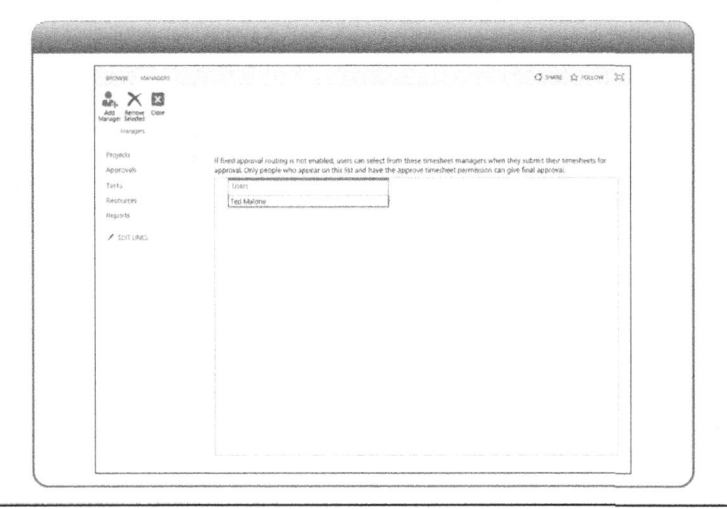

If fixed approval routing is not enabled, users can select from these timesheet managers when they submit their timesheets for approval. Only people who appear on this list and have the Approve Timesheet permission can give final timesheet approval.

How to Add a Timesheet Manager

1. On the **PWA Settings** page, in the **Time and Task Management** section, click **Timesheet Managers**.
2. Click **Add Manager**.
3. On the **Pick Resource** page, select the person you are adding as a manager. You can also use the Search box to help find the person in the list.
4. Click **OK**.

NOTE: To remove a manager, select the person in the list on the Timesheet Managers page, and then click **Remove Selected**.

Practice: Configuring Time & Task Management Settings

In this practice, you will:
- Configure Financial Periods
- Configure Time Reporting Periods
- Create a Line Classification
- Configure Timesheet Settings and Defaults
- Create Categories for Administrative Time
- Configure Task Settings and Display

In this practice, you will configure settings that control how timesheets and task updates behave and what resources can do with them. You will set up timesheet classifications and categories for different types of administrative time. You will also configure the task settings, which includes the tracking method to be used for all project managers.

Exercise 1: Configure Financial Periods

In this exercise, you will define the fiscal year and the fiscal reporting periods.

1. On the **PWA Home** page, click **Settings** and click **PWA Settings**.
2. On the **PWA Settings** page, under **Time and Task Management** section, click **Fiscal Periods**.
3. On the **Fiscal Periods** page, in the **Manage Fiscal Period** section, select **<current year> - Undefined** and click **Define**.
4. On the **Define Fiscal Year Parameters** page, configure the following settings:

Setting	Perform the following:
Define Fiscal Period Start Date	Click on the calendar and select **1/1/<currentyear>**
Set Fiscal Year Creation Model	Select **Standard calendar year**
Define Period Naming Convention	In the **Prefix** box, type **M** In the **Suffix** box, type **-<current year>** *For example, the sample could display M1-2017*

5. Click **Create and Save**.
6. On the **Fiscal Periods** page, click **Save**.

Exercise 2: Configure Time Reporting Periods

In this exercise, you will create and configure the time reporting periods.

1. On the **PWA Settings** page, in the **Time and Task Management** section, click **Time Reporting Periods**.
2. On the **Timesheet Periods** page, configure the following settings:

Define Bulk Period Parameters section	
Setting	Perform the following:
Number of periods to be created:	Type **52**
Date the first period starts:	Select the first Sunday in January of the current year
Type the standard period length (days)	Type **7**

Define Batch Naming Convention section	
Setting	Perform the following:
Prefix	Type **W**
Suffix	Type **-<current year>**
	*For example, the sample could display **W1-2017***
	Click **Create Bulk**
Create Periods	Change the status of the timesheet periods to **Closed** prior to today's date

3. Click **Save**.

Exercise 3: Create a Line Classification

In this exercise, you will create a classification for tracking a new type of billable time.

4. On the **PWA Settings** page, under **Time and Task Management** section, click **Line Classifications**.
5. On the **Line Classifications** page, in the menu bar, click **New Classification**.
6. In the **Name** box, type **Operations**, in the Description box, type **Operational Work** and click **Save**.

Exercise 4: Configure Timesheet Settings and Defaults

In this exercise, you will configure the timesheet settings that control what appears on timesheets and how resources can fill them in.

1. On the **PWA Settings** page, under **Time and Task Management** section, click **Timesheet Settings and Defaults**.
2. On the **Settings and Defaults** page, configure the following settings:

Setting	Perform the following:
Project Web App Display	Select **The timesheet will use standard Overtime and Non-Billable time tracking** check box (default)
Default Timesheet Creation Mode	Select **Current task assignments** (default)
Timesheet Grid Column Units	Select **Days** (default)
Default Reporting Units	Select **Hours** (default) Standard timesheet day is **8** Standard timesheet work week is **40**
Hourly Reporting Limits	Accept the default, (**Maximum Hours per Timesheet is 999, Minimum Hours per Timesheet is 0 and Maximum Hours per Day is 999**)
Timesheet Policies	Clear **Allow future time reporting** check box. Clear **Allow new personal tasks** check box. Clear **Allow top-level time reporting** check box. **Task Status Manager Approval** Select **Enabled** and select **Require line approval before timesheet approval** check box
Auditing	Select **Enable Timesheet Auditing** check box
Approval Routing	Clear **Fixed Approval Routing** check box
Single Entry Mode	Select **Single Entry Mode** check box

3. Click **Save**

Exercise 5: Create Categories for Administrative Time

In this exercise, you will create a series of categories for tracking administrative time.

1. On the **PWA Settings** page, under **Time and Task Management** section, click **Administrative Time**.

2. On the **Administrative Time** page, in the menu bar, click **New Category**, complete the new category with the following settings:

Category	Status	Work Type	Approve	Always Display
Training	Open	Working	Yes	Selected
Travel	Open	Working	No	Cleared
Maternity Leave	Open	Non-Work	Yes	Cleared
Bereavement	Open	Non-Work	Yes	Cleared

3. Click **Save**.

Exercise 6: Configure Task Settings and Display

In this exercise, you will choose the update method for tasks and restrictions for task updates.

1. On the **PWA Settings** page, under **Time and Task Management** section, click **Task Settings and Display**.

2. On the **Task Settings and Display** page, complete the configuration using the following settings:

Setting	Perform the following:
Tracking Method	*Because we have selected Single Entry Mode, the Tracking Method is automatically selected and cannot be modified.*
Reporting Display	Select **Resources should report their hours worked every day** (default)
	In the **Week starts on** list, select **Sunday**
Protect User Updates	Clear **Allow users to define custom periods for task updates** check box
Define Near Future Planning Window	Accept the default (**2**)

3. Click **Save**.

Lesson 2: Configuring Operational Policies

- Configuring Additional Server Settings
- Active Directory Resource Pool Synchronization
- Configuring Connected SharePoint Sites

In this lesson, you will learn how to configure the operation policies of Project Online.

Objectives

After completing this lesson, you will be able to:

- Configure Additional Server Settings
- Configure Active Directory Resource Pool Synchronization
- Configure Connected Project Sites

Configuring Additional Server Settings

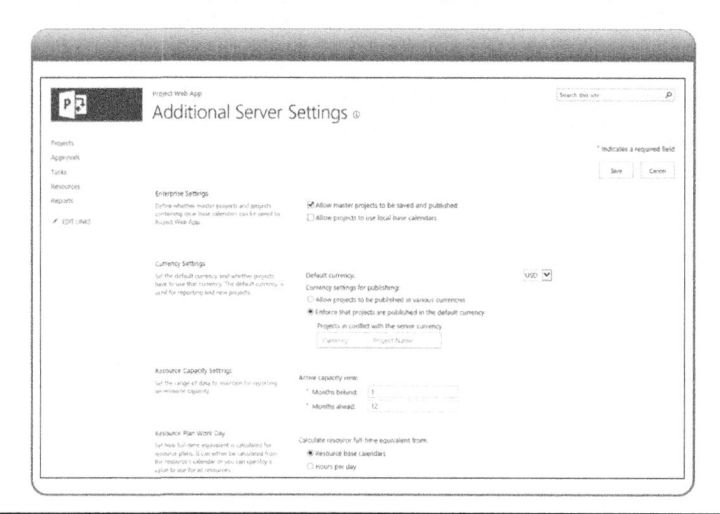

Project Professional Versions

Specify the version (build number) of Project Professional that is allowed to connect to Project Web App. Versions older than the specified version will be blocked from connecting to Project Web App.

Enterprise Settings

Define whether or not master projects and projects containing local base calendars may be published to the server.

Master Projects – Projects made up of two or more regular projects, called subprojects, which are often used to represent programs or collections of projects for reporting, analysis, and management purposes.

Local base calendars – Project, resource, and task calendars defined individually for each project.

- ☑ Allow master projects to be saved and published
- ☐ Allow projects to use local base calendars

Currency Settings

Use these settings to set the default server currency and whether or not projects may be published only in that currency.

Default currency: USD
Currency settings for publishing:
- ⦿ Allow projects to be published in various currencies
- ○ Enforce that projects are published in the default currency

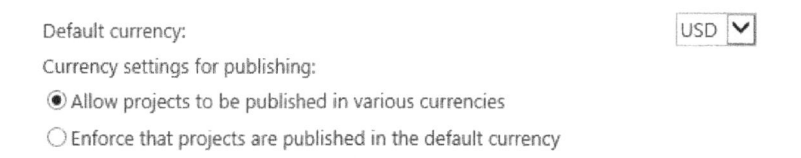

NOTE: *If your organization has different regions using different currencies, you should deploy a separate instance of Project Online for each region with its own currency setting.*

Resource Capacity Settings

Set the number of months behind and ahead of capacity data that the report database maintains for Resources. Set the schedule for maintaining forward-looking capacity data.

Active capacity view:
* Months behind: 1
* Months ahead: 12

IMPORTANT: *Increasing this setting will increase the amount of data stored per resource in the reporting table of the project database.*

Resource Plan Work Day

Enter the average length of the work day for all resources in a resource plan. The resource plan owner can use this average for all resources in the plan or use the individual resource calendar settings to calculate availability.

Calculate resource full-time equivalent from:
- ◉ Resource base calendars
- ○ Hours per day

Task Mode Settings

Define the default task mode and whether **Manually Scheduled tasks can be published to team members** on the server.

☑ Manually Scheduled tasks can be published to team members
Default task mode in new projects:
- ◉ Manually Scheduled
- ○ Automatically Scheduled

☑ Users can override default in Project Professional

Notification Email Settings

Determine whether users can subscribe to alerts and reminders.

Active Directory Enterprise Resource Pool Synchronization

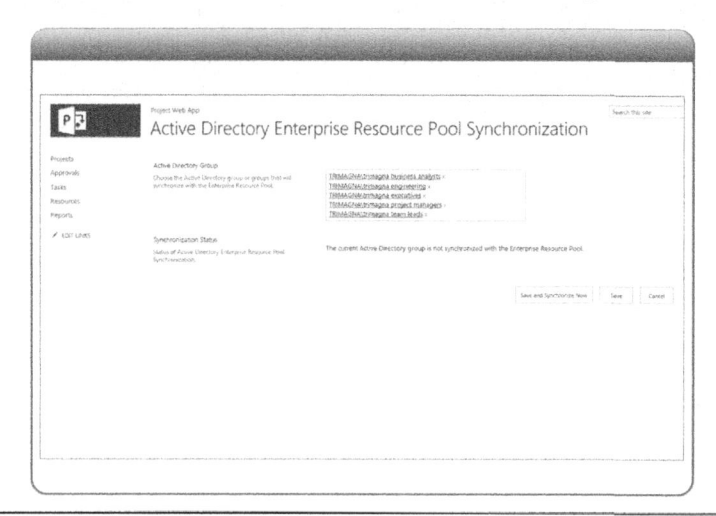

Project Online administrators can create/update resources in the Enterprise Resource Pool by using Active Directory Resource Pool Synchronization. This tool allows administrators to create the resources by mapping an Active Directory group directly with the Enterprise Resource Pool.

Active Directory Group

Type the name of an Active Directory group that is used to synchronize with the Enterprise Resource Pool.

Synchronization Status

Displays the status of the last Active Directory resource pool synchronization.

NOTE: You can synchronize up to five Active Directory groups with the Enterprise Resource Pool in Project Online. Each Active Directory group can contain nested groups whose members will also be synchronized.

Configuring Project Site Creation

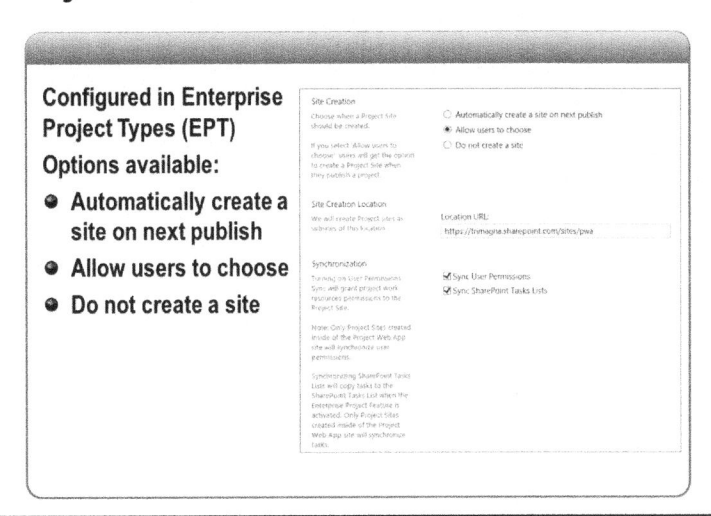

Configured in Enterprise Project Types (EPT)

Options available:

- Automatically create a site on next publish
- Allow users to choose
- Do not create a site

A project site contains issues, risks, documents, and deliverables associated with an enterprise project type (EPT). Depending on how the Project Site settings are configured, when a project manager creates and publishes a new project, an associated project site can either:

- Automatically create a site on next publish
- Allow users to choose
- Do not create a site

To Configure Project Site Settings

1. On the **PWA Settings** page, under **Workflow and Project Detail Pages** section, click **Enterprise Project Types**.

2. On the **Enterprise Project Types** page, click on the name of an EPT.

3. On the **<EPT Name>** page, under the **Site Creation** section, you can select either:

    a. **Automatically create a site on next publish**

    b. **Allow users to choose** – Project Managers will get the option to decide if they want to create a Project Site (or not) when they publish a project.

    c. **Do not create a site (default)**

   *NOTE: If you select either **Automatically create a site on next publish** or **Allow users to choose**, you can determine the location where project sites will be created for this EPT.*

4. Under the **Site Creation Location** section, you can change the location where project sites will be created. The default URL is:
   https://<companyname>.sharepoint.com/sites/pwa

5. Under the Synchronization section, there are two options available:

    a. **Sync User Permissions** – Turning this option on will grant project work resources permissions to the Project Site.

> ***NOTE:*** *Only Project Sites created inside of the Project Web App site will synchronize user permissions.*

b. **Sync SharePoint Tasks Lists** – Turning this option will copy tasks to the SharePoint Tasks List when the Enterprise Project Feature is activated.

> ***NOTE:*** *Only Project Sites created inside of the Project Web App site will synchronize tasks and SharePoint Tasks list will be read-only.*

Configuring Connected SharePoint Sites

A project site contains issues, risks, documents, and deliverables associated with an enterprise project. Depending on how the Project Site settings are configured, each time a project manager creates a new project, an associated project site is automatically created, or you can allow them to manually create a project site.

Create Site

This option manually creates a project site for the selected project if it was not created when the project was originally published. You can view the Project Sites list on the Connect SharePoint Sites page to determine whether or not a project has an existing site. A project without a project site will have no corresponding URL next to it in the Site Address column.

Edit Site Address

Select the project name or site address to edit the destination URL for a project site to point to a new site address. Changing the site address information breaks the existing link between the project and the existing project site.

Synchronize

Synchronizes users, permissions and other Project Online–related information between Project Online and SharePoint Online.

NOTE: The Synchronize setting is only available in Project Permission Mode. This setting is not available in SharePoint Permission Mode.

Delete Site

Permanently deletes the selected project workspace and all of its content.

Go to Project Site Settings

Opens the project site's site settings page where the sites administration settings are located. From the Site Settings page, you can make changes to the site, such as add or remove users, add Web Parts to the site, customize the site's look and feel, and many others.

Practice: Configuring Operational Policies

In this practice, you will:
- Configure Additional Server Settings
- Configure Enterprise Resource Pool Synchronization
- Configure Project Site Creation

In this practice, you will configure the Additional Server Settings and also synchronize a security group in Office 365 to the enterprise resource pool in Project Online.

Exercise 1: Configure Additional Server Settings

In this exercise, you will configure the additional server settings for Project Online.

1. On the **PWA Settings** page, under **Operational Policies** section, click **Additional Server Settings**.
2. On the **Additional Server Settings** page, complete the configuration using the following settings:

Enterprise Settings section	
Setting	**Perform the following:**
Allow master projects to be saved and published	Select check box (default)
Allow projects to use local base calendars	Clear check box (default)

Currency Settings section	
Setting	**Perform the following:**
Default server currency	Select **USD** (default)
Currency settings for publishing	Select **Enforce that projects are published in the server currency**

Resource Capacity Settings section	
Setting	**Perform the following:**
Months behind	Type **1** (default)

Module 4: Configuring Project Online

| Months ahead | Type **12** (default) |

Resource Plan Work Day section

Setting	Perform the following:
Calculate resource full-time equivalent from	Select **Resource base calendars** (default)

Task Mode Settings section

Setting	Perform the following:
Manually Scheduled Tasks can be published to team members	Select check box (default)
Default task mode in new projects	Select **Automatically Scheduled**
Users can override default in Project Professional	Select check box (default)

Notification Email Settings section

Setting	Perform the following:
Turn on notifications	Select check box

3. Click **Save**.

Exercise 2: Configure Enterprise Resource Pool Synchronization

In this exercise, you will synchronize the Team Members Grp security group in Office365 with the enterprise resource pool.

1. On the **PWA Settings** page, under **Operational Policies** section, click **Active Directory Resource Pool Synchronization**.

2. On the **Active Directory Enterprise Resource Pool Synchronization** page, in the **Active Directory Group** box, type **Team Members Grp** and click **Save and Synchronize Now**.

3. On the **PWA Settings** page, under **Operational Policies** section, click **Active Directory Resource Pool Synchronization**.

4. On the **Active Directory Enterprise Resource Pool Synchronization** page, the **Synchronization Status** should display the following message:

   The Enterprise Resource Pool was successfully synchronized on...

5. Click **Cancel**.

Exercise 3: Configure Project Site Creation

In this exercise, you will configure the project site creation settings for the Enterprise Project EPT.

1. On the **PWA Settings** page, under **Workflow and Project Detail Pages** section, click **Enterprise Project Types**.
2. On the **Enterprise Project Types** page, click on **Enterprise Project**.
3. On the **Enterprise Project** page, under the **Site Creation** section, select **Allow users to choose**.
4. Under the **Site Creation Location** section, accept the default URL.
5. Under the **Synchronization** section, select the check boxes for **Sync User Permissions** and **Syncs SharePoint Tasks Lists**.
6. At the bottom of the page, click **Save**.

Lesson 3: Importing Resources and Project Plans

- Overview of Importing Data
- Importing Resources to Enterprise
- Importing Projects to Enterprise

Resource and project plans can be imported to Project Online in many ways. The method you chose depends on the location and data type. In this lesson, you will learn how to choose the correct method to import your organizations resources and also the tasks to be completed prior to importing.

Objectives

After completing this lesson, you will be able to:

- Import Resources using the Import Resources to Enterprise Wizard
- Import Projects using the Import Projects to Enterprise Wizard

Overview of Importing Data

- **Use Project Online Professional to Import Data using:**
  - Import Project to Enterprise
  - Import Resources to Enterprise
- **Must be logged in to Project Online**

One of the key activities in deploying Microsoft Project Online is importing resources and projects into PWA. There are several ways to import data into Project Online.

Import Resources to Enterprise

One method of importing resources is to use the Import Resources to Enterprise wizard. This is the easiest way to get a group of resources into the system particularly if the list of resources is already in a project file that is being used as a central resource pool. Using the wizard, it is possible to point to project files and import the resources within them to the enterprise pool. This wizard can be used even if you do not have a single central pool file. You can run it as many times as you like against as many files as you have that contain resource data. It is more complex to import several smaller files, however, as the risk of duplicate resources increases. You can also import resources by using the Import Projects to Enterprise wizard.

Import Projects to Enterprise

If you use the Import Projects to Enterprise wizard to import specific projects, a screen displays the list of local resources within that file. It offers an option to map a resource to an existing enterprise resource or to import it into the enterprise resource pool.

Importing Resources to Enterprise

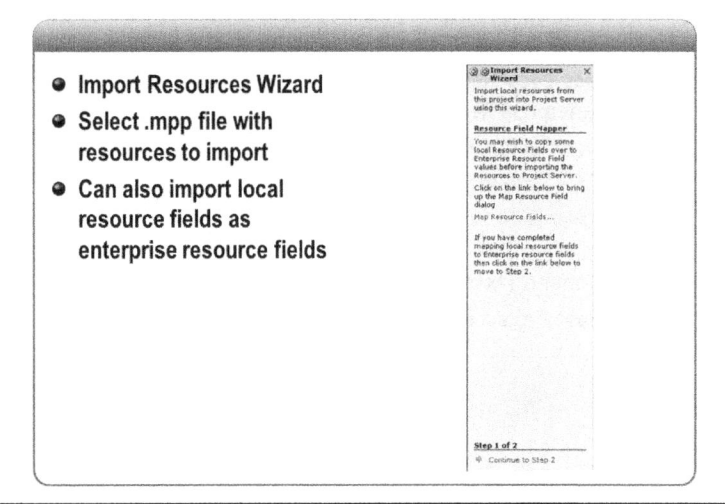

- Import Resources Wizard
- Select .mpp file with resources to import
- Can also import local resource fields as enterprise resource fields

Use the following procedure to run the Import Resources Wizard. Before you run the wizard, make sure you have created a single project plan in which to import your resources.

How to Import Resources into the Enterprise Resource Pool

1. In Microsoft Project Professional, click the **Resources** tab, click **Add Resources** and click **Import Resources to Enterprise**.

2. In the **Open** dialog box, navigate to the location of the .mpp file where the resources you want imported are located and click **Open**.

3. In the **Import Resources Wizard** pane, if you have any local resource fields that you might want copied to an enterprise resource field, click the **Map Resource Fields...** link, otherwise, click **Continue to Step 2**.

4. In the **Import Resources Wizard – Confirm Resources** pane, verify that there are no errors. If there errors you must correct the error by double clicking the resource name, make the corrections in the **Resource Information** dialog box and then click **Validate Resources**.

5. Once there are no errors, click **Save and Finish**.

Importing Projects to Enterprise

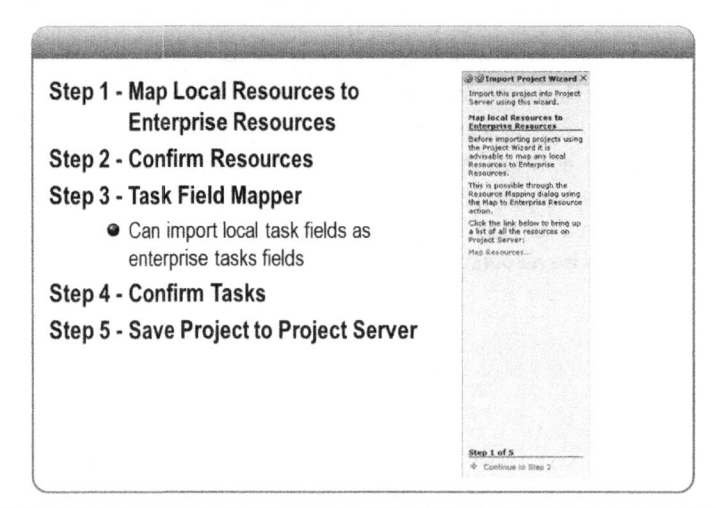

Step 1 - Map Local Resources to Enterprise Resources
Step 2 - Confirm Resources
Step 3 - Task Field Mapper
- Can import local task fields as enterprise tasks fields

Step 4 - Confirm Tasks
Step 5 - Save Project to Project Server

The following steps describe how to use the Import Project Wizard in Project Online Professional to import project plans to Project Online.

1. In Microsoft Project Professional, from the **File** tab, click **Open**.

2. In the **Open** dialog box, navigate to the location of the .mpp file where the project you want imported is located and click **Open**.

3. From the **File** tab, click **Save As**, select **Use Import Wizard** check box and click **Save**.

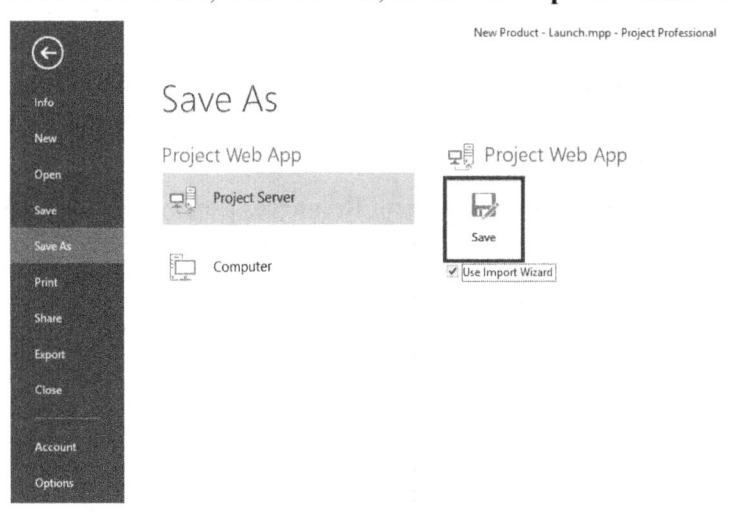

4. In the **Import Project Wizard** pane, if you have any local resources assigned to the project you need to map the local resources to enterprise resources, click **Map Resources...** link.

5. In the **Map Local onto Enterprise Resources** dialog box, in the **Map to Enterprise** column you can select one of two choices and click **OK**:

    No, keep local – This action keeps the resource in the project but it will not be an enterprise resource. This should be done only for resources that will likely never be used on any other projects.

Yes – This action allows you to map a resource in your plan to a specified resource in the Enterprise Pool. If the name in the imported plan matches exactly with a resource name in the enterprise resource pool, this action is selected for you by default.

6. In the **Import Project Wizard** pane, click **Continue to Step 2**.

7. In the **Import Resources Wizard – Confirm Resources** pane, verify that there are no errors. If there are errors you must correct the error by double clicking the resource name, make the corrections in the **Resource Information** dialog box and then click **Validate Resources**. If there are no errors, click **Continue to Step 3**.

8. In the **Import Resources Wizard – Map Task Custom Fields** pane, if you have any task fields that you might want imported to an enterprise task field, click the **Map Fields...** link, otherwise, click **Continue to Step 4**.

9. In the **Import Resources Wizard – Confirm Tasks** pane, verify that there are no errors. If there are errors you must correct the error by double clicking the resource name, make the corrections in the **Task Information** dialog box and then click **Continue to Step 5**.

10. In the **Import Project Wizard – Complete Import** pane, click the **Save** link.

11. In the **Save to Project Web App** dialog box, complete the following fields and click **Save**:

Setting	Perform the following:
Name	Type the name of the project
Type	Select **Project** or **Template**
Calendar	Select a predefined enterprise calendar
Custom Fields	Complete any custom project fields

12. In the **Import Project Wizard – Complete Import** pane, click **Save and Finish**.

NOTE: Saving the project to Project Online does not make it available for viewing in Project Web App – Project Center. To view the project plan, you must first publish the project.

Practice: Importing Resources and Projects

In this practice, you will:
- Import Resources
- Import Projects
- Manually Create a Project Site

In this practice, you will import resources and projects from .MPP files into Project Online.

Exercise 1: Import Resources

In this exercise, you will import two .MPP files that only contain resources into the enterprise resource pool.

1. Launch **Microsoft Project Online Professional** and in the **Login** dialog box, select **Project Online** and click **OK**.
2. On the **Project Recent** page, in the navigation pane, click **Open Other Projects**.
3. On the **Open** page, click **This PC** and then click **Browse**.
4. In the **Open** dialog box, navigate to the location of the course files and double click **EntResources.mpp**.
5. From the **Resource** tab, click **Add Resources → Import Resources to Enterprise**.
6. In the **Import Resources Wizard** pane, click **Continue to Step 2**.
7. In the **Import Resources Wizard – Confirm Resources** pane, change the **Import** settings to **Yes** for all resources, except for **Guy Gilbert** and click **Save and Finish**.
8. From the **File** tab, click **Open**.
9. On the **Open** page, click **This PC** and then click **Browse**.
10. In the **Open** dialog box, navigate to the location of the course files and double click **Generic Resources.mpp**.
11. From the **Resource** tab, click **Add Resources → Import Resources to Enterprise**.
12. In the **Import Resources Wizard** pane, click **Continue to Step 2**.
13. In the **Import Resources Wizard – Confirm Resources** pane, change the **Import** settings to **Yes** for all resources and click **Save and Finish**.
14. Switch to **Internet Explorer**.
15. On the **PWA Settings** page, in the quick launch menu, click **Resources**.

16. Switch back to **Project Online Professional** and close the MPP files without saving them.

Exercise 2: Import Projects

In this exercise, you will import local projects into Project Online.

1. In the **Project Professional** navigation pane, click **Open**.
2. On the **Open** page, click **This PC** and then click **Browse**.
3. In the **Open** dialog box, navigate to the location of the course files and double click **New Product - Develop Prototype.mpp**.
4. From the **Project** tab, click **Project Information**.
5. Change the **Start date** to 1 month from today's date and click **OK**.
6. From the **File** menu, click **Save As**, select **Use Import Wizard** check box and then click **Save**, as shown below:

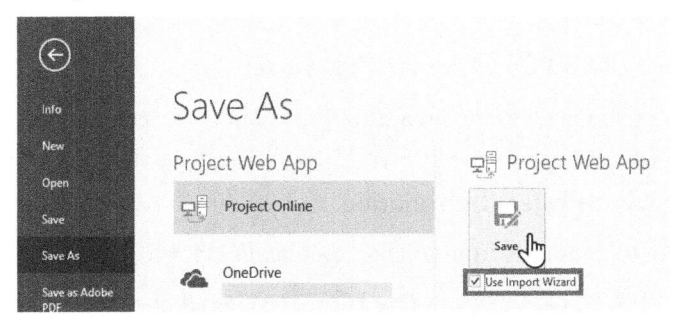

7. In the **Import Project** pane, click the **Map Resources...** link
8. In the **Map Local to Enterprise Resources** dialog box, for each local resource listed, make the following mappings and click **OK**.

Local Resource	Map to Enterprise?	Enterprise Resource:
Beth	Yes	Beth Melton
Cheryl	Yes	Cheryl McGuire
Eric	Yes	Eric Denekamp
Jerry	Yes	Jerry Smith
Management	Yes	_EXECUTIVE
Zeshan	Yes	Zeshan Sattar

   Notice the Resource Names have changed in the Gantt View

9. In the **Import Project** pane, click **Continue to Step 2**.
10. In the **Import Project – Import Resources** pane, verify that there are no errors and click **Continue to Step 3**.
11. In the **Import Project – Map Task Custom Fields** pane, click **Continue to Step 4**.
12. In the **Import Project – Confirm Tasks** pane, verify that there are no errors and click **Continue to Step 5**.

13. In the **Import Project Wizard – Complete Import** pane, click the **Save** link, as shown below:

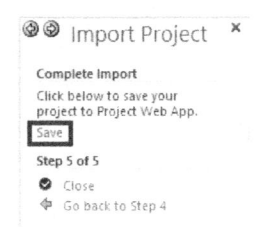

14. In the **Save to Project Web App** dialog box, in the **Name** box, type **New Product – Develop Prototype** and click **Save**.

15. In the **Import Project Wizard – Complete Import** pane, click **Close**.

16. From the **File** tab, on the **Info** page, click **Publish**.

17. In the **Publish Project: New Product – Develop Prototype** dialog box, select **Do not create a site at this time** and click **Publish**.

18. From the **File** tab, click **Open**.

19. On the **Open** page, click **This PC** and then click **Browse**.

20. In the **Open** dialog box, navigate to the location of the course files and double click **New Product - Launch.mpp**.

21. From the **Project** tab, click **Project Information**.

22. Change the **Start date** to 2 months from today's date and click **OK**.

23. From the **File** menu, click **Save As**, select **Use Import Wizard** check box and then click **Save**.

24. In the **Import Project** pane, click the **Map Resources...** link.

25. In the **Map Local to Enterprise Resources** dialog box, for each local resource listed, make the following mappings and click **OK**:

Local Resource	Map to Enterprise?	Enterprise Resource:
Engineering	Yes	_ENGINEERING
Field Service	Yes	_Logistics
Manufacturing	Yes	_MANUFACTURING
Marketing	Yes	_Marketing
Product Manager	Yes	Christophe Fiessenger
Product Support	Yes	_Product Support
Sales	Yes	_Sales

Notice the Resource Names have changed in the Gantt View

26. In the **Import Project** pane, click **Continue to Step 2**.

27. In the **Import Project – Import Resources** pane, verify that there are no errors and click **Continue to Step 3**.

28. In the **Import Project – Map Task Custom Fields** pane, click **Continue to Step 4**.

29. In the **Import Project – Confirm Tasks** pane, verify that there are no errors and click **Continue to Step 5**.

30. In the **Import Project Wizard – Complete Import** pane, click the **Save** link.

31. In the **Save to Project Web App** dialog box, in the **Name** box, type **New Product - Launch** and click **Save**.

32. In the **Import Project Wizard – Complete Import** pane, click **Close**.

33. From the **File** tab, on the **Info** page, click **Publish**.

34. In the **Publish Project: New Product - Launch** dialog box, select **Create a site for this project** and click **Publish**.

35. Close **Project Online Professional** and check in and save all projects.

36. Switch to **Internet Explorer** and from the **Resource Center** page, in the Quick Launch, click **Projects**. *The imported projects should be listed.*

Exercise 3: Manually Create a Project Site

In this exercise, you will create a project site for import local projects into Project Online.

1. On the **Project Center** page, click **Settings**, click **PWA Settings**.

2. On the **PWA Settings** page, under **Operational Policies** section, click **Connected SharePoint Sites**.

3. On the **Connected SharePoint Sites** page, select **New Product – Develop Prototype**, in the menu bar, click **CREATE SITE**.

4. On the **Create Project Site** dialog box, accept the defaults and click **OK**.

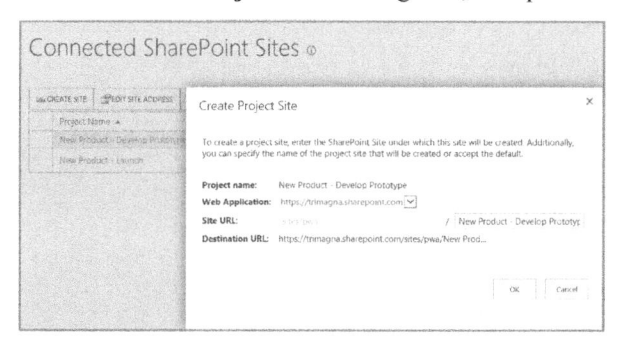

5. On the **Connected SharePoint Sites** page, click the **Site Address** for **New Product – Develop Prototype**. The project site should be available.

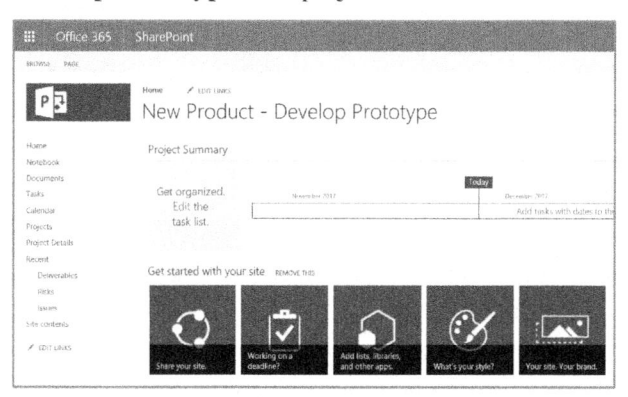

Summary

> **In this module, you learned how to:**
> - Configure Time & Task Management Settings
> - Configure Operational Policies
> - Import Resources and Project Plans

In this module, you learned how to configure time and task settings that include timesheets and the corresponding settings. You also learned how to configure Project Online operational policies and how to import resources and project plans.

Objectives

After completing this module, you learned how to:

- Configure Time & Task Management Settings
- Configure Operational Policies
- Import Resources and Project Plans

Module 5: Configuring Enterprise Data Settings

Contents

Module Overview 1
Lesson 1: Configuring Enterprise Custom Fields 2
    Overview of Custom Fields 3
    Enterprise Custom Fields 4
    Creating Enterprise Custom Fields 5
    Creating Lookup Tables 8
    Preconfigured Enterprise Custom Fields 10
    Practice: Configuring Enterprise Custom Fields 11
Lesson 2: Configuring Enterprise Objects 17
    Resource Breakdown Structure (RBS) 18
    Security Rules and the RBS 19
    Configuring Enterprise Calendars 20
    Configuring the Enterprise Global Template 21
    Practice: Configuring Enterprise Objects 22
Summary 25

EXCLUSIVELY PUBLISHED BY

PMO Logistics
679 Roberta Avenue
Winnipeg, Manitoba, Canada R2K 0K9

Copyright © 2017 by Roland Perreaux

All rights reserved. No part of the contents of this document may be reproduced or transmitted in any form or by any means without written permission of the publisher.

Information in this document, including URL and other Internet Web site references, is subject to change without notice. Unless otherwise noted, the example companies, organizations, products, domain names, e-mail addresses, logos, people, places, and events depicted herein are fictitious, and no association with any real company, organization, product, domain name, e-mail address, logo, person, place, or event is intended or should be inferred. Complying with all applicable copyright laws is the responsibility of the user. Without limiting the rights under copyright, no part of this document may be reproduced, stored in or introduced into a retrieval system, or transmitted in any form or by any means (electronic, mechanical, photocopying, recording, or otherwise), or for any purpose, without the express written permission of PMO Logistics Inc.

The names of manufacturers, products, or URLs are provided for informational purposes only and PMO Logistics makes no representations and warranties, either expressed, implied, or statutory, regarding these manufacturers or the use of the products with any Microsoft technologies. The inclusion of a manufacturer or product does not imply endorsement of Microsoft of the manufacturer or product. Links are provided to third party sites. Such sites are not under the control of PMO Logistics and PMO Logistics is not responsible for the contents of any linked site or any link contained in a linked site, or any changes or updates to such sites. PMO Logistics is not responsible for webcasting or any other form of transmission received from any linked site. PMO Logistics is providing these links to you only as a convenience, and the inclusion of any link does not imply endorsement of PMO Logistics of the site or the products contained therein.

PMO Logistics may have patents, patent applications, trademarks, copyrights, or other intellectual property rights covering subject matter in this document. Except as expressly provided in any written license agreement from PMO Logistics, the furnishing of this document does not give you any license to these patents, trademarks, copyrights, or other intellectual property.

PMO Logistics, Professional Training Series, Upgrading Skills Series, TriMagna Corporation and TriMagna Corporation logo are either registered trademarks or trademarks of PMO Logistics Inc. in Canada, the United States and/or other countries.

Microsoft, Active Directory, Internet Explorer, Outlook, Project Server, SharePoint, SQL Server, Visual Studio, Windows and Windows Server are either registered trademarks or trademarks of Microsoft Corporation in the United States and/or other countries.

All other trademarks are property of their respective owners.

Author: Rolly Perreaux, PMP, MCSE, MCT

Publisher: PMO Logistics
Developmental Editor: Heather Perreaux
Cover Graphic Design: Andrea Ardiles
Technical Testing: Underground Studioworks

Post-Publication:
Errata List Contributors:

Module Overview

- Configuring Enterprise Custom Fields
- Configuring Enterprise Objects

In this module, you will learn how to create enterprise custom fields in order to add structure to your project, resource and task data. You will learn how to configure enterprise objects, such as the RBS, Enterprise Global and Enterprise Calendars.

Objectives

After completing this module, you will be able to:

- Configure Enterprise Custom Fields
- Configure Enterprise Objects

Lesson 1: Configuring Enterprise Custom Fields

- Overview of Custom Fields
- Enterprise Custom Fields
- Creating Enterprise Custom Fields
- Creating Lookup Tables
- Preconfigured Enterprise Custom Fields

In this lesson, you will learn about enterprise custom fields. You will also learn about enterprise custom field definitions and how to create enterprise custom fields and lookup tables.

Objectives

After completing this lesson, you will be able to:

- Understand Custom Fields
- Create Enterprise Custom Fields
    - Enterprise Resource Fields
    - Enterprise Project Fields
    - Enterprise Task Fields
- Create Lookup Tables

Overview of Custom Fields

- **Custom Fields allow you to:**
  - Write your own formulas to be calculated
  - Create a list of values
  - Display a graphical indicator instead of the actual data.
  - Create a hierarchical structure of custom fields

- **Two types of Custom Fields**
  - Local Custom Fields
  - Enterprise Custom Fields

A field is a location in a sheet, form, or chart that contains specific information about a task, resource, or assignment. For example, in a sheet, each column is a field.

Custom fields contain customized data such as, text, flags, numbers, dates, cost, start and finish dates, and durations. You can customize these fields to obtain the information you want using formulas, specific value calculations, or graphical indicators.

With custom fields, you can:

- Write your own formulas, including references to other fields, to be calculated in a custom field.
- Create a list of values for a custom field to ensure fast and accurate data entry.
- Display a graphical indicator in a custom field instead of the actual data.
- Create a hierarchical structure of custom fields for information in your project.

There are two types of custom fields:

Local Custom Fields – Used by a project manager within the scope of a particular project or personal standards.

Enterprise Custom Fields – Used by a project manager or project management office (PMO) to collect data for rollup reporting across the organization. For enterprise task and project custom fields, Project Online supports the scoping to a specific program (collection of projects), so that an enterprise custom field can be defined that applies to a subset of projects.

Enterprise Custom Fields

- **Typically created and defined to provide project management standards within an organization**
- **Used to:**
    - Organize
    - Search
    - Create custom views of the data
- **Created at the project, task, and resource level in Project Web App**

You can use Enterprise Custom Fields to create a set of project management standards that can be applied across your organization and to enhance the capabilities of Project Online.

You can create Enterprise Custom Fields at the project, task and resource levels. It is important to determine which specific Enterprise Custom Fields your organization needs when you review your business requirements while planning your Project Online deployment. It is best to do this after you have performed a gap analysis by comparing the capabilities of Project Server against the business needs of your organization.

For example, a group of executives in an organization wants to be able to view project data by division. To achieve this you would need to define a new custom field to identify the various divisions within the organization.

Similarly, an organization might require custom ways to define the resources within each division. If project managers need to be able to select resources for their projects based on location, then they need an easy way to identify the location for each resource. Enterprise Resource Outline Codes can be used to help project managers determine availability based on location; however, sometimes additional identifiers are required.

The most important use for Enterprise Custom Fields is to enable organizations to enforce consistency across all projects.

The key thing to remember is that the Project Online database is just a large database collection of projects, tasks and resources. To help you organize, search and generate reports, use enterprise custom fields to define the attributes for projects, tasks and resources.

Creating Enterprise Custom Fields

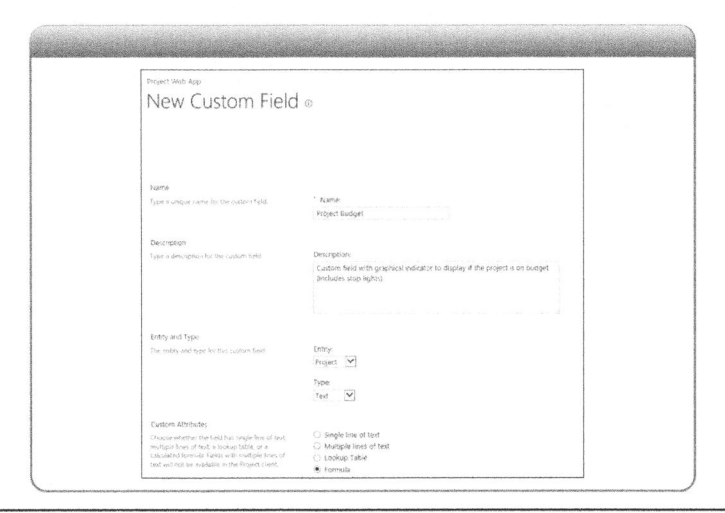

Field Name

Project Online allows you to define an unlimited number of enterprise custom fields providing that custom field names:

- Are unique
- Do not have special characters (! @ # $ % ^ & *)
- Have a maximum length of 50 characters.

NOTE: You cannot change the entity context or data type after you have saved a given enterprise custom field definition. You must create a new enterprise custom field to change the entity context or type.

Field Entities

Each enterprise custom field must be defined with a field entity. The following field entities determine the type of data attributes that will be required dependent on the selection:

- Project
- Resource
- Task

Field Types

Each enterprise custom field must be defined with a field type. The following field types determine the type of data that will be stored:

- **Cost** – Data consists of monetary units that cannot have any free-form text. Cost values can be positive or negative.
- **Date** – Data consists of calendar values that conform to Windows operating system settings and within the Microsoft Project date range 1/1/1984 through 12/31/2049.
- **Duration** – Data consisting of Project duration values can have an optional suffix, such as "h" for hours, "d" for days, etc. Duration values can be positive or negative.

- **Flag** – Data consists of a Yes or No value only.
- **Number** – Data consists of numeric values and cannot contain free-form text.
- **Text** – Data consists of free-form text consisting of text, numbers, dates, etc. to a maximum of 256 characters.

Custom Attributes

Determines if the enterprise custom field has a lookup table, a calculated formula, or neither.

Lookup Table

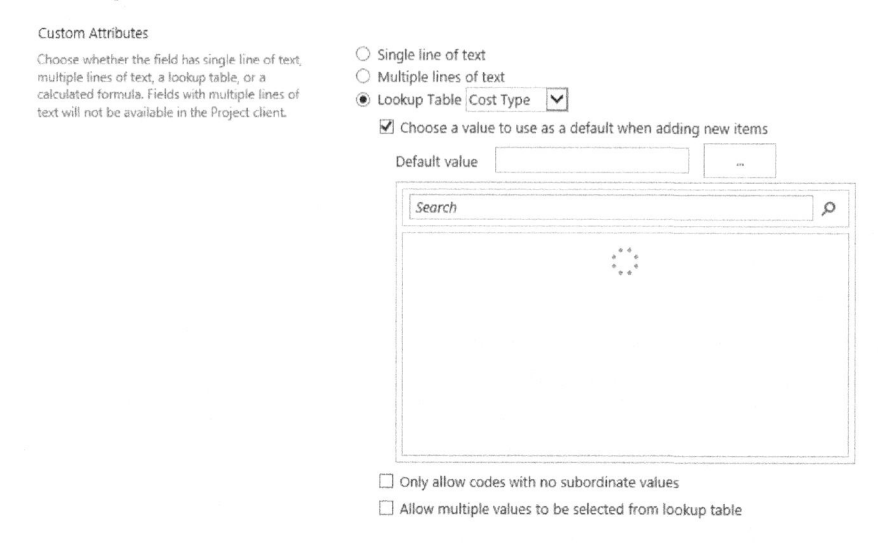

Formula

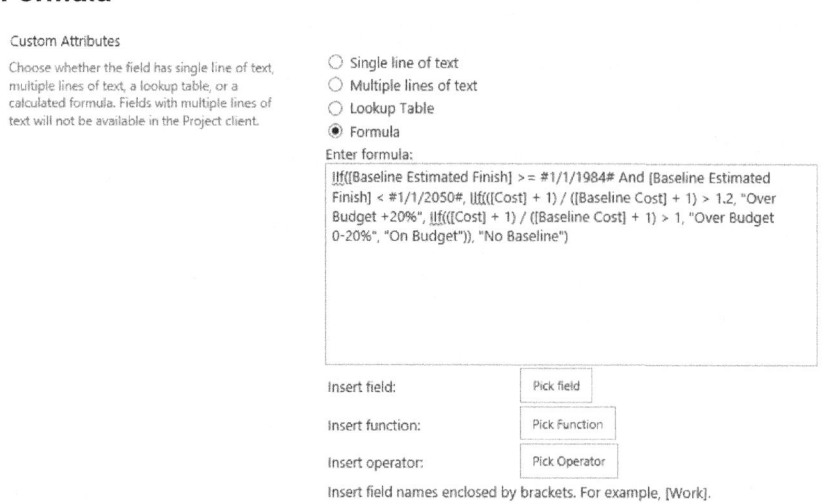

Calculation for Summary Rows

You can choose how this field gets rolled up for summary rows including grouping summaries. Available only if Resource or Task entity is selected.

Calculation for Assignment Rows

You can choose how this field gets rolled down to assignments. Available only if Resource or Task entity is selected.

Values to Display

Choose whether you want just the data or graphical indicators to be displayed.

Values to Display

Choose whether you want just the data or graphical indicators to be displayed. Graphical indicators are not displayed in all areas of Project Web App.

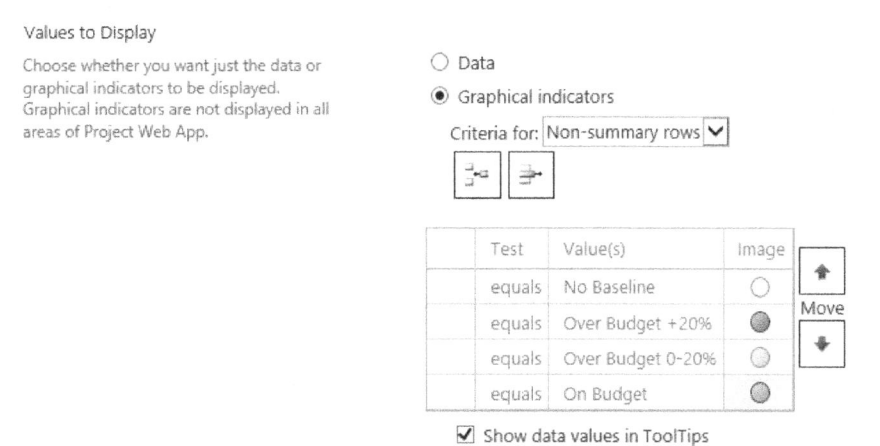

○ Data
◉ Graphical indicators
Criteria for: Non-summary rows

Test	Value(s)	Image
equals	No Baseline	○
equals	Over Budget +20%	●
equals	Over Budget 0-20%	○
equals	On Budget	○

☑ Show data values in ToolTips

Required

You can make this field a required field. Since required fields can increase work for users, carefully consider whether or not to make this a required field.

Creating Lookup Tables

The organization's business requirements might specify only standardized data values to be allowed within projects, resources, or tasks. You can define lookup tables to establish these standards with restrictive values.

Lookup Table Name

Project Online allows you to define an unlimited number of lookup tables providing that custom field names:

- Are unique
- Do not have special characters (! @ # $ % ^ & *)
- Have a maximum length of 50 characters.

Lookup Table Types

Each enterprise custom field must be defined with a field type. The following field types determine the type of data that will be stored:

- **Cost** – Data consists of monetary units that cannot have any free-form text. Cost values can be positive or negative.
- **Date** – Data consists of calendar values that conform to Windows operating system settings and within the Microsoft Project date range 1/1/1984 through 12/31/2049.
- **Duration** – Data consists of Project duration values can have an optional suffix, such as "h" for hours, "d" for days, etc. Duration values can be positive or negative.
- **Number** – Data consists of numeric values and cannot contain free-form text.
- **Text** – Data consists of free-form text consisting of text, numbers, dates, etc. to a maximum of 256 characters.

Code Mask

The code mask defines each level with the number and sequence of characters that are allowed. It is only available by selecting **Text** type. The Code Preview box displays the structure of your lookup field codes.

Code preview:

AAA.000000.****

Code mask:

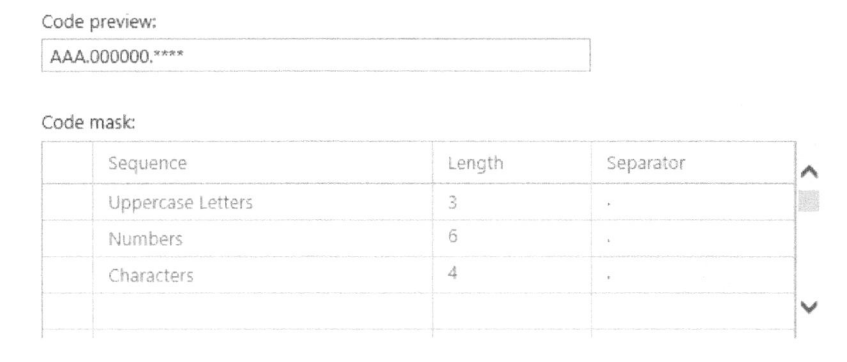

Sequence	Length	Separator
Uppercase Letters	3	.
Numbers	6	.
Characters	4	.

Lookup Table

In the Lookup Table section, type values in the table and create a hierarchical relationship between each value. The values must conform to the code mask definition that you specified in the Code Mask section.

These are the values from which the users can select when they enter data into the custom fields. To help create the value's hierarchical relationship, there are a series of action icons that allow you to perform actions such as Cut, Copy, Paste, Insert, Delete, Outdent, Indent, Expand, Collapse, Expand All, Export to Excel and Print.

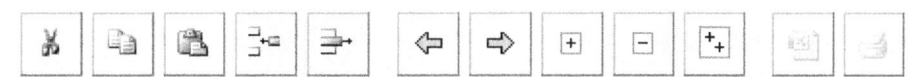

Move rows within the table using the up and down arrows.

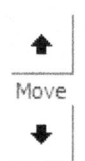

Sort rows in the table by row number (using the table structure you created), or sorted alphabetically in ascending or descending order.

Display order for lookup table:
- ◉ By row number
- ○ Sort ascending
- ○ Sort descending

Preconfigured Enterprise Custom Fields

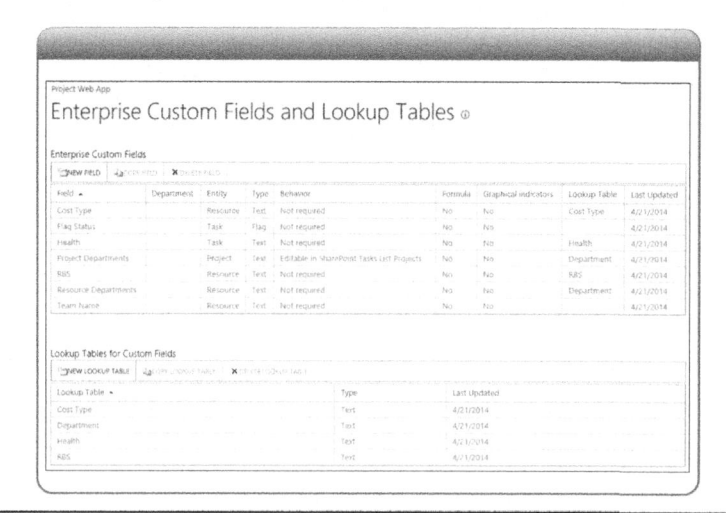

When installing Project Online, by default it creates the following preconfigured enterprise custom fields:

Cost Type – A custom resource field with a corresponding lookup table that is not defined.

Flag Status – A custom task field.

Health – A custom task field with a corresponding text lookup table with the following entries:

- Not Specified
- On schedule
- Late
- Early
- Blocked
- Completed

Project Departments – A custom project field associated with Departments lookup table that is not defined.

RBS – A custom resource field with a corresponding lookup table that is not defined.

Resource Departments – A custom resource field associated with Departments lookup table that is not defined.

Team Name – A custom resource field without a defined lookup table. Used for assigning Team Resource Pool.

Practice: Configuring Enterprise Custom Fields

In this practice, you will:
- Create Custom Enterprise Lookup Table
- Create Custom Enterprise Fields
  - Resource Field
  - Project Fields

In this practice, you will create a custom enterprise lookup table and custom enterprise fields. You will then restore saved custom fields and lookup tables and configure the enterprise data settings for Project Online.

Exercise 1: Create Custom Enterprise Lookup Table

In this exercise, you will create a series of lookup tables and enterprise custom resource fields that will be assigned to resources in the enterprise resource pool.

1. From Internet Explorer, click **Settings** and then click **PWA Settings**.

2. On the **PWA Settings** page, under **Enterprise Data** section, click **Enterprise Custom Fields and Lookup Tables**.

3. On the **Enterprise Custom Fields and Lookup Tables** page, under **Lookup Tables for Custom Fields** section, click **New Lookup Table**.

4. On the **New Lookup Table** page, in the **Name** box, type **Project Type Table**, in the **Type** box, select **Text**.

5. Under the **Code Mask** section, accept the default.

6. Under the **Lookup Table** section, add the following values:
   - DevOps
   - Governance
   - IT
   - Marketing
   - Operations
   - R&D

7. Click **Save**.

8. On the **Enterprise Custom Fields and Lookup Tables** page, under **Lookup Tables for Custom Fields** section, click **New Lookup Table**.

9. On the **New Lookup Table** page, in the **Name** box, type **Skills Table**, in the **Type** box, select **Text**.

10. Under the **Code Mask** section, in the second row, complete the following settings:

Column	Perform the following:
Sequence	Select **Characters**
Length	Select **Any**
Separator	Select **.** (period)

11. Repeat step 5 three more times until you have 4 levels

    For example, the **Code Preview** displays *.*.*.*

12. Open **File Explorer** and navigate to the location where the course files are saved and open **RESOURCE ENTERPRISE FIELD - SKILLS.txt** and copy the contents of the file.

    SHORTCUT: Press Ctrl +A keys to select all and then Ctrl +C keys to copy.

13. Switch back to **Internet Explorer** and under the **Lookup Table** section, under the **Value** column, right click in the first row and click **Paste**.

14. In the **Internet Explorer** dialog box, click **Allow access**.

15. In the lookup table, in the empty column before the **Level** column, click on the second row that has the value of **Advertising**, (this will highlight the entire row).

16. Hold down the **Shift** key and click on the empty column on the row that has the value of **Web Content**, then release **Shift** key.

    This will highlight all the rows in grey.

17. On the **Lookup Table** menu bar, click **indent** (right arrow).

    This indents all the highlighted rows to Level 2 of **COMMUNICATIONS**.

Level	Value	Description
1	COMMUNICATIONS	
2	Advertising	
2	Electronic	
2	Letters	
2	Presentations	
2	Proposals	
2	Sales	
2	Telecommunications	
2	Web Content	
1	ENGINEERING	
1	Chemical	

18. Switch to **File Explorer** and open **RESOURCE SKILLS HIERARCHY.txt**

Module 5: Configuring Enterprise Data Settings

19. Using the highlighting and indenting technique, complete the lookup table based on the file.

   > **IMPORTANT:** *PWA will not allow you to save the lookup table if you have any spaces before or after the value.*
   >
   > *Make sure to delete all empty rows at the bottom of the lookup table before saving the lookup table, or else you will generate the following error message and you will need to restart the entire procedure.*
   >
   > ***The lookup table could not be saved due to the following reason(s):***
   > ***An unknown error has occurred.***

20. In the lookup table, scroll down to the bottom of the lookup table and delete all empty rows.

21. Review one last time and click **Save**.

Exercise 2: Create Custom Enterprise Fields

In this exercise, you will create an enterprise custom resource field and project fields.

1. On the **Enterprise Custom Fields and Lookup Tables page**, under the **Enterprise Custom Fields** section, click **New Field**.

2. In the **New Custom Field** page, complete the configuration using the following settings:

Column	Perform the following:
Name:	Type **Skills**
Description:	Type **Custom resource field listing skills inventory**
Entity:	Select **Resource**
Type:	Select **Text**
Custom Attributes:	Select **Lookup Table**, in the list select **Skills Table**.
	Select **Only allow codes with no subordinate values**.
	Select **Allow multiple values to be selected from lookup table**.
	Select **Use this field for matching generic resources**

3. Click **Save**.

4. On the **Enterprise Custom Fields and Lookup Tables page**, under the **Enterprise Custom Fields** section, click **New Field**.

5. In the **New Custom Field** page, complete the configuration using the following settings:

Column	Perform the following:
Name:	Type **Project Budget**
Description:	Type **Custom field with graphical indicator to display if the project is on budget (includes stop lights)**
Entity:	Select **Project**

Type:	Select **Text**
Custom Attributes:	Select **Formula**
	Switch to **File Explorer** and open **PROJECT BUDGET-FORMULA.txt** and copy the contents of the file.
	Switch back to **Internet Explorer**, right click in the **Formula** box and click **Paste**.
Values to Display	Select **Graphical indicators**
	In the **Criteria for list**, select **Non-summary rows**.
	1. Under the **Test** column, in the first row, click **Insert row**.
	2. Under the **Test** column, in the first row, select **equals**.
	3. Switch to **File Explorer**, open **PROJECT BUDGET-INDICATORS.txt** and copy the first row.
	4. Under the **Value(s)** column, right click and click **Paste**.
	5. Under the **Image** column, select the corresponding image:

Test	Value(s)	Image
equals	No Baseline	White Stoplight
equals	Over Budget +20%	Red Stoplight
equals	Over Budget 0-20%	Yellow Stoplight
equals	On Budget	Green Stoplight

6. Repeat Steps 2-5 for each value, as shown below:

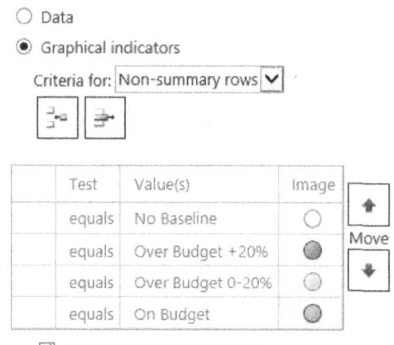

7. Select **Show data values in ToolTips** check box.

6. Click **Save**.

7. On the **Enterprise Custom Fields and Lookup Tables page**, under the **Enterprise Custom Fields** section, click **New Field**.

8. In the **New Custom Field** page, complete the configuration using the following settings:

Column	Perform the following:
Name:	Type **Project Schedule**

Module 5: Configuring Enterprise Data Settings 5-15

Description:	Type **Custom field with graphical indicator to display if the project is on time (includes stop lights)**
Entity:	Select **Project**
Type:	Select **Text**
Custom Attributes:	Select **Formula**
	Switch to **File Explorer** and open **PROJECT SCHEDULE-FORMULA.txt** and copy the contents of the file.
	Switch back to **Internet Explorer**, right click in the **Formula** box and click **Paste**.
Values to Display	Select **Graphical indicators**
	In the **Criteria for list**, select **Non-summary rows**.
	1. Under the **Test** column, in the first row, click **Insert row**.
	2. Under the **Test** column, in the first row, select **equals**.
	3. Switch to **File Explorer**, open **PROJECT SCHEDULE-INDICATORS.txt** and copy the first row.
	4. Under the **Value(s)** column, right click and click **Paste**.
	5. Under the **Image** column, select the corresponding image:

Test	Value(s)	Image
equals	No Baseline	White Stoplight
equals	Late +5 Days	Red Stoplight
equals	Late 1-5 Days	Yellow Stoplight
equals	On Schedule	Green Stoplight

6. Repeat Steps 2-5 for each value, as shown below:

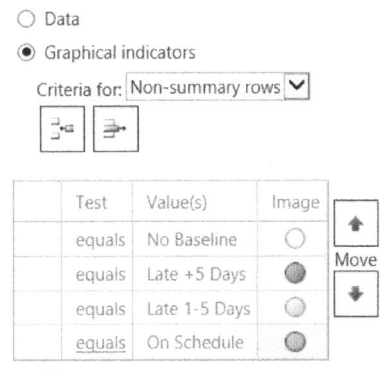

○ Data
◉ Graphical indicators
Criteria for: Non-summary rows ▼

Test	Value(s)	Image
equals	No Baseline	○
equals	Late +5 Days	●
equals	Late 1-5 Days	○
equals	On Schedule	●

☑ Show data values in ToolTips

7. Select **Show data values in ToolTips** check box.

9. Click **Save**.

10. On the **Enterprise Custom Fields and Lookup Tables page**, under the **Enterprise Custom Fields** section, click **New Field**.

11. In the **New Custom Field** page, complete the configuration using the following settings:

Column	Perform the following:
Name:	Type **Project Category**
Description:	Type **Custom field to display listing of project categories.**
Entity:	Select **Project**
Type:	Select **Text**
Custom Attributes:	Select **Lookup Table**, in the list, select **Project Category Table**.

12. Click **Save**.

13. Close the **File Explorer** and all **Notepad** windows.

Lesson 2: Configuring Enterprise Objects

- Resource Breakdown Structure (RBS)
- Security Rules and the RBS
- Configuring Enterprise Calendars
- Configuring the Enterprise Global Template

In this lesson, you will learn about the Resource Breakdown Structure (RBS) and the role it plays in providing additional security to Project Server. You will also learn how to configure Enterprise Calendars and the Enterprise Global Template.

Objectives

After completing this lesson, you will be able to:

- Configure a Resource Breakdown Structure
- Understand how security and the RBS works together
- Configure Enterprise Calendars
- Configure the Enterprise Global Template

Resource Breakdown Structure (RBS)

- **Hierarchical structure that represents an organization through an Enterprise Resource Field**
- **Based on:**
  - Organization Chart
  - Geographic Locations
  - Matrix organization
- **Used in the following features:**
  - Resource Substitution Wizard
  - Data Analysis
  - Build Team
  - Two default security categories:
    - My Direct Reports and My Resources

The RBS is a predefined custom resource field with a hierarchical structure that represents your enterprise resources. It allows you to create project plans with detailed resource assignments and to compare this workload with detailed resource availabilities. The RBS also enables roll-up of both resource assignments and availability data to a higher level.

Use RBS to define the reporting relationships among users and resources in your organization. Project Online uses the relationships that are defined in the RBS to simplify the management of access for users and groups. This is an integral component of resource management and application security.

Most organizations define their RBS using one of the following three scenarios:

- Organization-based RBS
- Geographic-based RBS
- Matrix organization-based RBS

The RBS custom resource field is used in the following Project Online features:

- Resource Substitution Wizard
- Build Team from Enterprise in Project Online Professional
- Build Team in Project Web App
- Two default security categories: My Direct Reports and My Resources

NOTE: You can define only one RBS custom resource field for your organization

Security Rules and the RBS

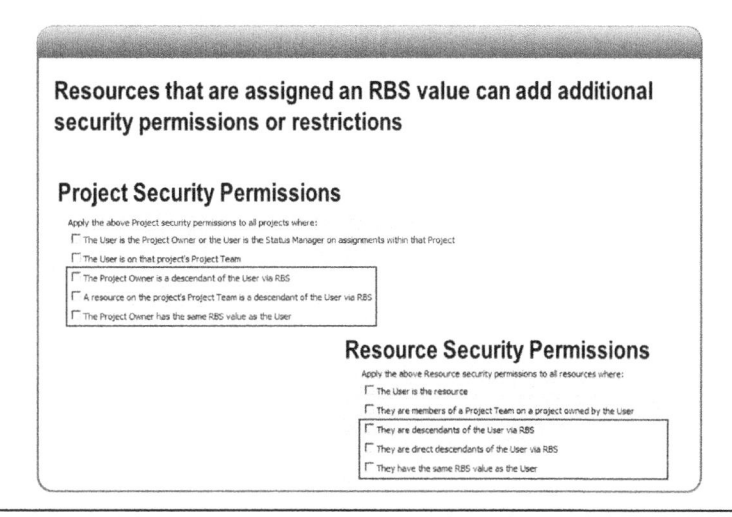

> Resources that are assigned an RBS value can add additional security permissions or restrictions
>
> **Project Security Permissions**
> Apply the above Project security permissions to all projects where:
> ☐ The User is the Project Owner or the User is the Status Manager on assignments within that Project
> ☐ The User is on that project's Project Team
> ┌───┐
> │ ☐ The Project Owner is a descendant of the User via RBS   │
> │ ☐ A resource on the project's Project Team is a descendant of the User via RBS │
> │ ☐ The Project Owner has the same RBS value as the User   │
> └───┘
>
> **Resource Security Permissions**
> Apply the above Resource security permissions to all resources where:
> ☐ The User is the resource
> ☐ They are members of a Project Team on a project owned by the User
> ┌───┐
> │ ☐ They are descendants of the User via RBS   │
> │ ☐ They are direct descendants of the User via RBS   │
> │ ☐ They have the same RBS value as the User   │
> └───┘

The Project Online security model uses the RBS to determine enterprise resource hierarchy. This hierarchy is used by the category security settings that allow members of a category to view resources that are their descendants (they manage) or direct descendants (they manage directly), or to view projects and resources managed by the resources that are their descendants. The managing versus managed relationship is based on where the resource is located in the RBS hierarchy. There are six security category settings that directly relate to the RBS.

Project Security Permissions

Apply the above Project security permissions to all projects where:

☐ The User is the Project Owner or the User is the Status Manager on assignments within that Project

☐ The User is on that project's Project Team

☐ The Project Owner is a descendant of the User via RBS

☐ A resource on the project's Project Team is a descendant of the User via RBS

☐ The Project Owner has the same RBS value as the User

Resource Security Permissions

Apply the above Resource security permissions to all resources where:

☐ The User is the resource

☐ They are members of a Project Team on a project owned by the User

☐ They are descendants of the User via RBS

☐ They are direct descendants of the User via RBS

☐ They have the same RBS value as the User

Configuring Enterprise Calendars

- **Used to determine working time for all projects, tasks, and resources**
- **Four Types of Enterprise Calendars:**
  - Base Calendars
  - Project Calendars
  - Resource Calendars
  - Task Calendars

Enterprise Calendars are the primary scheduling mechanism used to determine working time for all projects, tasks, and resources.

Project Online includes four types of calendars:

Base Calendars – Used for two purposes, either directly as a task or project calendar or as a template for resource calendars. The Standard Enterprise Calendar is the only base calendar that is created and assigned by default to projects and resources. It is based on a traditional Monday-Friday, 8:00 A.M. to 5:00 P.M. work schedule with a one hour break.

Project Calendar – Used as the base calendar to designate the default work schedule for all tasks in a project.

Resource Calendar – Used to reflect specific working hours, vacations, leaves of absence, and planned personal time for individual resources. Individual resource calendars for enterprise resources are stored in the Enterprise Resource Pool.

Task Calendar – Used for situations in which you want to schedule a task outside of the normal working times defined by a project calendar or resource calendar.

You can add Enterprise Calendars to the Enterprise Global template in the following ways:

- Use Organizer in Project Online Professional to copy a calendar from an existing project into the Enterprise Global template.
- By creating them manually in PWA under Server Settings, Enterprise Calendars.

All Enterprise Calendars are stored in the Enterprise Global template. When a user launches Project Online Professional and connects to PWA, the latest Enterprise Calendars are cached on the local computer as part of the cached Enterprise Global template. This enables a user to use Enterprise Calendars when working on a project in offline mode.

Configuring the Enterprise Global Template

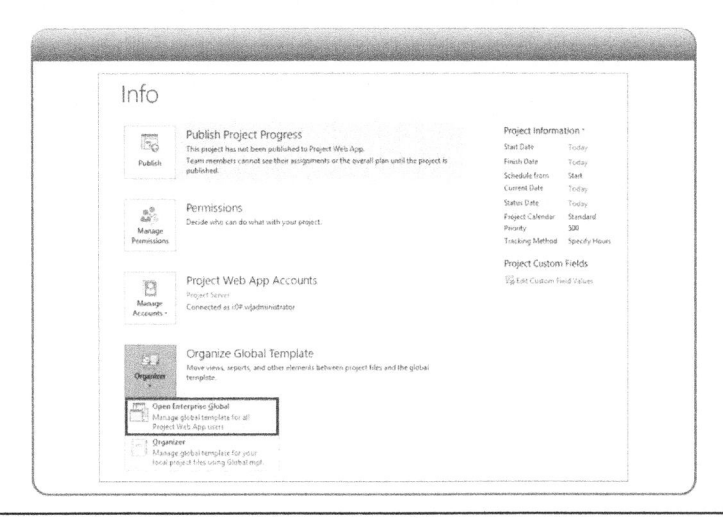

The Enterprise Global is a template with a collection of default settings, such as views, tables, macros and custom fields that are used by all projects across the organization. The Enterprise Global template ensures that all projects within an organization adhere to standards.

How to Edit the Enterprise Global Template

1. Launch Project Online Professional, from the **File** tab, click **Info**, in the **Organize Global Template** section, click **Organizer** and click **Open Enterprise Global**.

    The Enterprise Global template is checked out and opened as a new project.

2. Make all the necessary changes as required.

    REMEMBER: *You are making the changes for all PWA users*

3. From the **File** tab, click **Close**.

4. In the **Close** dialog box, select **Save**, select **Check-in** and click **OK**.

NOTE: *Restart Project Professional to view the changes.*

Practice: Configuring Enterprise Objects

In this practice, you will:
- Modify Standard Enterprise Calendar
- Create a New Enterprise Calendar
- Customize the Enterprise Global

Exercise 1: Modify Standard Enterprise Calendar

In this exercise, you will modify the standard calendar to include non-working days (holidays).

1. From Internet Explorer, click **Settings** and then click **PWA Settings**.

2. On the **PWA Settings** page, under the **Enterprise Data** section, click **Enterprise Calendars**.

3. On the **Enterprise Calendars** page, select **Standard,** in the menu bar, click **Edit Calendar**.

   *This will launch **Project Online Professional**.*

4. In the **Change Working Time** dialog box, on the calendar, navigate to **January** and select **January 1**, on the **Exceptions** tab, in the **Name** box, type **New Year's Day** and press **Enter**.

5. Click on **New Year's Day** and click **Details**.

6. In the **Details for New Year's Day** dialog box, configure the following settings and click **OK**:

Setting	Perform the following:
Recurrence pattern	Select **Yearly**
Range of recurrence	Select **End after**:
	In the **occurrences** box, type **20**

   IMPORTANT: *Remember to select the calendar date of the holiday, prior to typing the holiday name.*

7. (OPTIONAL) Configure the remaining US holidays as non-working days for 20 occurrences using the following list.

Holiday Name	Recurrence Pattern

Module 5: Configuring Enterprise Data Settings

Martin L King's Birthday	Yearly - Third Monday in January
President's Day	Yearly - Third Monday in February
Memorial Day	Yearly - Last Monday in May
Independence Day	Yearly - July 4
Labor Day	Yearly - First Monday in September
Columbus Day	Yearly - Second Monday in October
Veteran's Day	Yearly - November 11
Thanksgiving Day	Yearly - Fourth Thursday in November
Christmas Day	Yearly - December 25

8. When complete, click **OK** and close **Project Online Professional** without saving.

NOTE: When switching back to PWA, Internet Explorer may not respond. Close and re-launch Internet Explorer and open PWA again to Enterprise Calendars.

Exercise 2: Create a New Enterprise Calendar

In this exercise, you will create a custom enterprise calendar based on the Standard calendar.

1. On the **Enterprise Calendars** page, select **Standard** and in the menu bar, click **Copy**.
2. On the **Copy Calendar** page, in the **Name** box, type **Evening Shift** and click **OK**.
3. On the **Enterprise Calendars** page, select **Evening Shift** and in the menu bar, click **Edit**.

   Notice the holidays are included in the new calendar.

4. In the **Change Working Time** dialog box, click **Work Weeks** tab, select **[Default]** and click **Details**.
5. In the **Details for [Default]** dialog box, in the **Select day(s)** list, click **Monday**, hold the **Shift** key and click **Friday**, then select **Set day(s) to these specific working times**.
6. Configure the following settings and click **OK**:

From	To
4:00 PM	7:30 PM
8:00 PM	12:00 AM

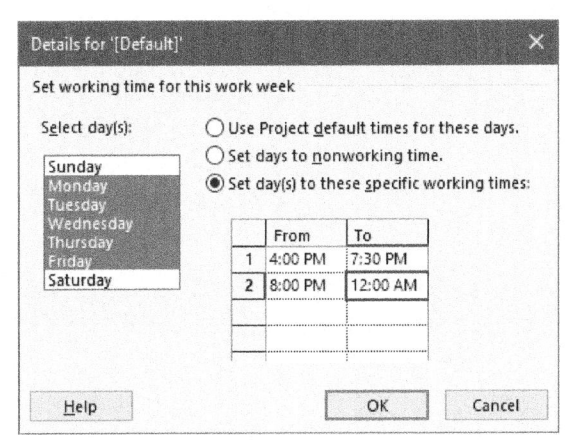

> *NOTE: If the **From** and **To** times do not set properly, try using military time instead. (16:00 to 19:30 and 20:00 to 0:00)*

7. In the **Change Working Time** dialog box, click **OK**.
8. Do not close **Project Online Professional**.

Exercise 3: Customize the Enterprise Global Template

In this exercise, you will customize the Enterprise Global Template.

1. From the **File** tab, click **Info**, in the **Organize Global Template** section, click **Organizer** and click **Open Enterprise Global**.
2. In the **Checked-out Enterprise Global - Microsoft Project** window, from the **File** tab, click **Options**.
3. In the **Project Options** window, in the navigation pane, click **General**, in the **Date format** box, select **Wed Jan 28, '09**.
4. In the navigation pane, click **Advanced** and configure the following settings:

General section	Select **Prompt for project info for new projects** check box
Display options for this project section	Select **All New Projects** Select **Show project summary task** check box

5. In the **Project Options** window, click **OK**.
6. From the **File** tab, click **Save**, then close **Project Online Professional**.
7. In the **Microsoft Project** dialog box, click **Yes**, to check-in the Enterprise Global.

Summary

In this module, you learned how to:
- Configure Enterprise Custom Fields
- Configure Enterprise Objects

In this module, you learned how to create enterprise custom fields in order to add structure to your project, resource and task data. You learned how to configure enterprise objects, such as the Resource Breakdown Structure, Enterprise Calendars and the Enterprise Global Template.

Objectives

After completing this module, you learned how to:

- Configure Enterprise Custom Fields
- Configure Enterprise Objects

This page is intentionally left blank

Module 6: Customizing Project Sites

Contents

Module Overview 1
Lesson 1: Working with Project Sites and Elements 2
  Working with Risks 3
  Working with Issues 4
  Working with Document Libraries 5
  Working with Deliverables 6
  Understanding Project Site Permissions 7
  Practice: Working with Project Sites 8
Lesson 2: Creating a Custom Project Site Template 9
  Customizing the Project Site 10
  Adding Web Parts to Project Site 11
  Adding SharePoint Apps 13
  Saving the Project Site as SharePoint Template 14
  Practice: Creating a Custom Project Site Template 15
Summary 19

EXCLUSIVELY PUBLISHED BY

PMO Logistics
679 Roberta Avenue
Winnipeg, Manitoba, Canada R2K 0K9

Copyright © 2017 by Roland Perreaux

All rights reserved. No part of the contents of this document may be reproduced or transmitted in any form or by any means without written permission of the publisher.

Information in this document, including URL and other Internet Web site references, is subject to change without notice. Unless otherwise noted, the example companies, organizations, products, domain names, e-mail addresses, logos, people, places, and events depicted herein are fictitious, and no association with any real company, organization, product, domain name, e-mail address, logo, person, place, or event is intended or should be inferred. Complying with all applicable copyright laws is the responsibility of the user. Without limiting the rights under copyright, no part of this document may be reproduced, stored in or introduced into a retrieval system, or transmitted in any form or by any means (electronic, mechanical, photocopying, recording, or otherwise), or for any purpose, without the express written permission of PMO Logistics Inc.

The names of manufacturers, products, or URLs are provided for informational purposes only and PMO Logistics makes no representations and warranties, either expressed, implied, or statutory, regarding these manufacturers or the use of the products with any Microsoft technologies. The inclusion of a manufacturer or product does not imply endorsement of Microsoft of the manufacturer or product. Links are provided to third party sites. Such sites are not under the control of PMO Logistics and PMO Logistics is not responsible for the contents of any linked site or any link contained in a linked site, or any changes or updates to such sites. PMO Logistics is not responsible for webcasting or any other form of transmission received from any linked site. PMO Logistics is providing these links to you only as a convenience, and the inclusion of any link does not imply endorsement of PMO Logistics of the site or the products contained therein.

PMO Logistics may have patents, patent applications, trademarks, copyrights, or other intellectual property rights covering subject matter in this document. Except as expressly provided in any written license agreement from PMO Logistics, the furnishing of this document does not give you any license to these patents, trademarks, copyrights, or other intellectual property.

PMO Logistics, Professional Training Series, Upgrading Skills Series, TriMagna Corporation and TriMagna Corporation logo are either registered trademarks or trademarks of PMO Logistics Inc. in Canada, the United States and/or other countries.

Microsoft, Active Directory, Internet Explorer, Outlook, Project Server, SharePoint, SQL Server, Visual Studio, Windows and Windows Server are either registered trademarks or trademarks of Microsoft Corporation in the United States and/or other countries.

All other trademarks are property of their respective owners.

Author: Rolly Perreaux, PMP, MCSE, MCT

Publisher: PMO Logistics
Developmental Editor: Heather Perreaux
Cover Graphic Design: Andrea Ardiles
Technical Testing: Underground Studioworks

Post-Publication:
Errata List Contributors:

Module Overview

- Working with Project Sites and Elements
- Creating a Custom Project Site Template

Project sites provide an effective and centralized platform for the communications, collaboration, and storage of various types of project information. If project sites are provisioned for programs and their projects, you can realize the benefits of publishing and sharing risks, issues, documents, deliverables, and other information in an integrated way with a project schedule.

In this module, you will learn how to work with project sites and their elements. You will also learn how to create a custom project site and then how to create a site template.

Objectives

After completing this module, you will be able to:

- Work with Project Sites and Elements
- Create a Custom Project Site Template

Lesson 1: Working with Project Sites and Elements

- Working with Risks
- Working with Issues
- Working with Document Libraries
- Working with Deliverables
- Understanding Project Site Permissions

A Project Site contains risks, issues, document libraries and deliverables that are associated with an enterprise project. As a Project Online administrator, you can configure Microsoft Project Web App so that each time a project manager publishes a new project, an associated project site is automatically created, or you can allow the project manager to manually create the project site.

SharePoint Online provides risk and issues tracking and a document library for each project published to Project Online. In addition to tracking risks, issues, and documents on a per-project basis, you can associate them with multiple projects or even with each other.

Objectives

After completing this lesson, you will be able to:

- Understand the following SharePoint elements:
    - Risks
    - Issues
    - Document Libraries
    - Deliverables
- Understand how Project Site permissions work

Working with Risks

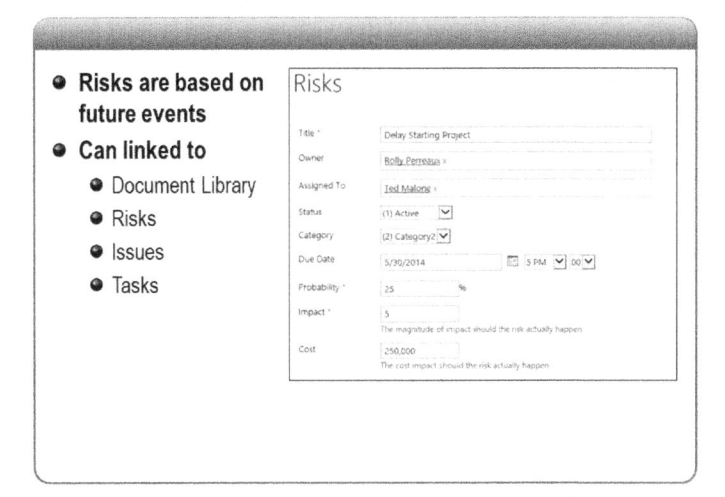

- **Risks are based on future events**
- **Can linked to**
  - Document Library
  - Risks
  - Issues
  - Tasks

Risks are events or conditions that can have a positive or negative effect on the outcome of a project. SharePoint Online integrates risk tracking into the overall project life cycle to help track events that can have adverse future effects on a project. In many ways, risk tracking is a fundamental part of project management. Project Web App has the ability to create and track risks for each published project.

Tracking Risks Can Include:

- Creating custom risks for any project
- Linking risks to multiple tasks or multiple projects
- Associating issues with risks
- Viewing and searching risk histories
- Receiving alerts as to the status of a risk
- Creating risk reports

Working with Issues

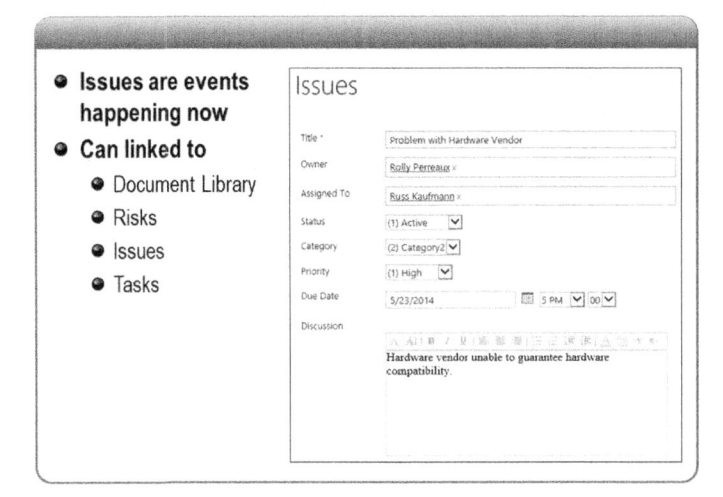

- **Issues are events happening now**
- **Can linked to**
  - Document Library
  - Risks
  - Issues
  - Tasks

Issue tracking is an integral part of project management. Issues are used to capture work items not identified in assignments or tasks within projects. Team members can submit issues about project details. Project managers, resource managers, and team leads can then determine if the issue will affect the project and identify a strategy to manage the issue before it becomes a major problem.

The conceptual difference between risks and issues are:

- Risks are potential events that might arise in the future
- Issues are identified events happening now

Tracking Issues can include:

- Creating custom issues for any project
- Linking issues to multiple tasks or multiple projects
- Associating risks with issues
- Viewing and searching issue histories
- Receiving alerts as to the status of an issue
- Creating issue reports

Working with Document Libraries

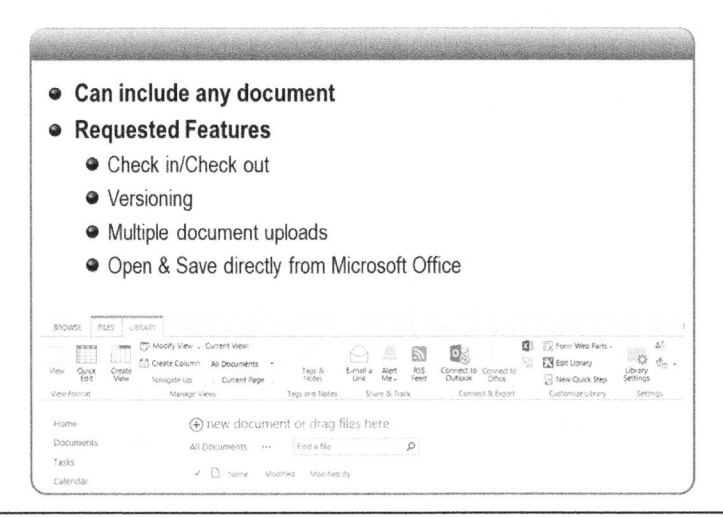

- **Can include any document**
- **Requested Features**
  - Check in/Check out
  - Versioning
  - Multiple document uploads
  - Open & Save directly from Microsoft Office

Document libraries are integrated into Project Online with the documents feature in SharePoint Online. Documents can be used to capture non-structured, project-related information that could not otherwise be included in a typical project.

Document Libraries can include:

- Associating documents with projects, tasks, risks, issues and deliverables
- Checking in, checking out, and versioning
- Saving documents directly from Microsoft Office applications and other applications
- Receiving e-mail notifications when the status of a document has changed

Working with Deliverables

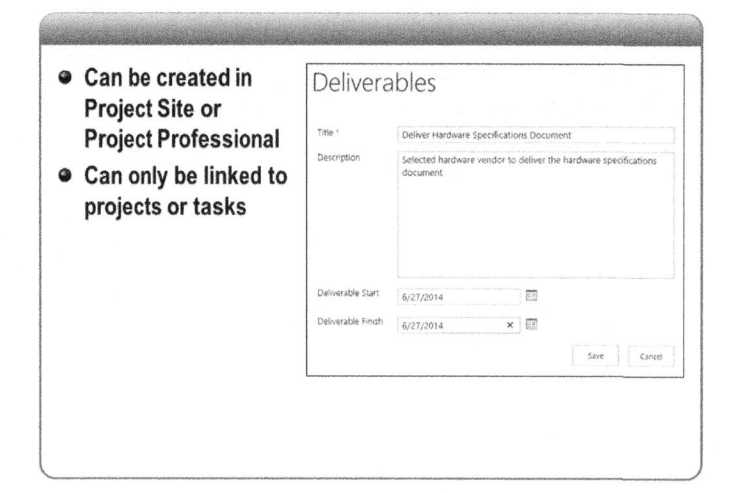

- **Can be created in Project Site or Project Professional**
- **Can only be linked to projects or tasks**

[Screenshot of Deliverables dialog with Title "Deliver Hardware Specifications Document", Description "Selected hardware vendor to deliver the hardware specifications document", Deliverable Start 6/27/2014, Deliverable Finish 6/27/2014]

A deliverable is a tangible or intangible object that is produced as a result of project execution. Typically, the project team and project stakeholders agree on the project deliverables before the project begins.

Clarifying the deliverables before the project work begins can help ensure that project outcomes meet all the stakeholders' expectations and that the goals of the project align with the larger business goals.

> **NOTE:** *Deliverables can be created in either the Project Site or in Project Online Professional. But only Project Online Professional can link the deliverable to a task in the project.*

Understanding Project Site Permissions

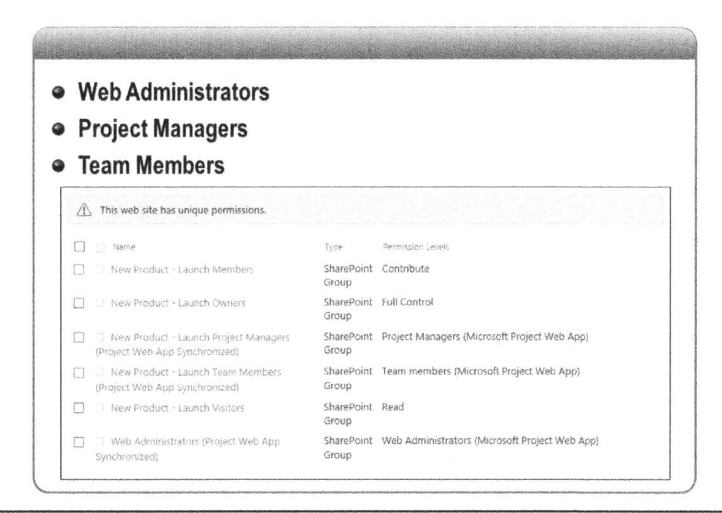

- **Web Administrators**
- **Project Managers**
- **Team Members**

Microsoft Project Online is completely dependent upon SharePoint Online to support its user interface, farm topology, and administration features. Security is tightly integrated between Project Online and SharePoint Online. When a project is published or if the Project Manager has to allow it, a project site is created. Users that have been added to the project or who have been granted permission to the project are added to at least one of three SharePoint Services groups:

- **Web Administrators (Project Web App Synchronized)** – Users who have Manage SharePoint permission in Project Online and are contributors to the project site. They can create and edit documents, issues, and risks.

- **Project Managers (Project Web App Synchronized)** – Users who have published this project or who have Save Project permission in Project Online and are contributors to the project site. They can create and edit documents, issues, and risks.

- **Team Members (Project Web App Synchronized)** – Users who have assignments in this project in Project Online and are contributors to the project site. They can create and edit documents, issues, and risks.

Project Online groups and SharePoint Online are synchronized whenever a project is published or the administrator selects a project site on the Project Sites page and then clicks Synchronize.

Practice: Working with Project Sites

In this practice, you will:
- Create and Publish a Temporary Project to create a Project Site

In this practice, you will create a temporary project in order to create a project site template.

Exercise 1: Create and Publish a Temporary Project

In this exercise, you will create and publish a temporary project to create a project site.

1. Launch **Project Online Professional** and login to **Project Online**.
2. On the **Project – Recent** page, click **Blank Project**.
3. In the **Project Information for 'Project1'** dialog box, click **Cancel**.
4. In the **Gantt Chart** view, in the **Task Name** column, underneath **Project1**, type **Task1** with duration of **1d**.
5. From the **File** tab, click **Save**.
6. On the **Save As** page, click **Project Online** and click **Save**.
7. In the **Save to Project Web App** dialog box, in the **Name** box, type **Admin Project** and click **Save**.
8. From the **File** tab, on the **Info** page, click **Publish**.
9. In the **Publish Project: Admin Project** dialog box, click **Create a site for this project** and click **Publish**.
10. Close **Project Online Professional** (make sure to save and check-in the project).
11. Switch to **Internet Explorer** and navigate to the **PWA Home** page.
12. On the **PWA Home** page, in the Quick Launch, click **Projects**.
13. On the **Project Center** page, click **Admin Project**.
14. From the **Project** tab, click **Project Site**.

    The Project Site for the Admin Project should be displayed.

Lesson 2: Creating a Custom Project Site Template

- Customizing the Project Site
- Adding Web Parts to the Project Site
- Adding SharePoint Apps
- Saving the Project Site as a SharePoint template
- Changing the default Project Site template

By default, only the Project Site template is used for the creation of future project sites. However, most organizations prefer to customize the default template or create a new template to be used that addresses their needs.

In this lesson, you will create a customized Project Site that will be used as a template for future Project Sites. You will save the site as a SharePoint Site Template and then modify the Project Site provisioning settings to use the new template as the basis for new Project Sites.

Objectives

After completing this lesson, you will be able to:

- Customize the Project Site
- Add Web Parts to the Project Site
- Add SharePoint Apps
- Save the Project Site as a SharePoint template
- Change the default Project Site template

Customizing the Project Site

> **Things to Consider:**
> - Versioning settings for Document Libraries, Risks and Issues
> - Check-in / Check Out settings
> - Document templates used in Document Libraries
> - Custom Lists
> - Look and Feel settings.
>   - Tree View
>   - Site Theme
>   - Top Link Bar
>   - Quick Launch
>
> **Always remember you are creating a standard**

Customizing the Project Site is simply a matter of deciding which web parts you want to add or remove as part of the site or the creation of new lists to be included. The main thing is that you customize the Project Site that you want so that it will become the template (the organization's standard) for all future Project Sites.

Considerations for Customizing

- Versioning settings for Document Libraries, Risks and Issues
- Check In / Check Out settings
- Document templates used in Document Libraries
- Custom Lists
- Look and Feel settings
    - Tree View
    - Site Theme
    - Top Link Bar
    - Quick Launch

IMPORTANT: *Do not modify the default columns used in Risks and Issues as they are used internally in Project Online. However, you can add additional columns.*

Adding Web Parts to Project Site

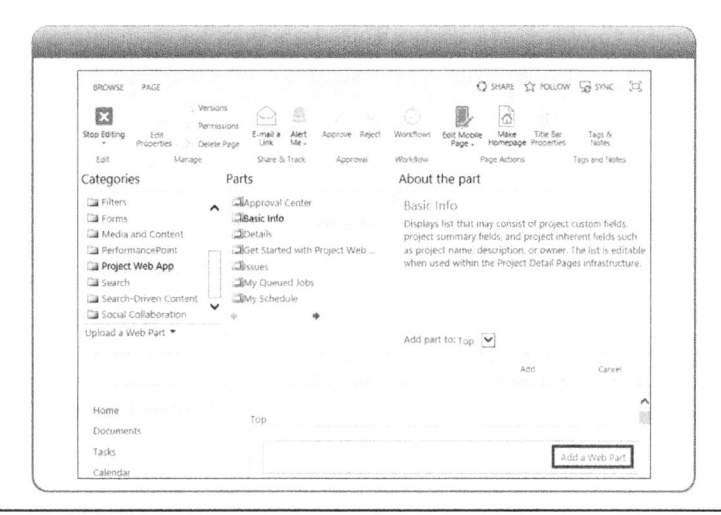

Web Parts are modular elements that add functionality to pages of SharePoint sites. The following is a list of common Web Part categories:

- **Apps** – Each app instance you have added to your site has an associated Web Part. The app Web Parts allows you to add a view into the data in your app to your web pages.

- **Blog** – Provides Web Parts for a blog site.

- **Business Data** – A group of Web Parts that displays business information, such as status, indicators, and other business data. This group also includes Web Parts for embedding Excel and Visio documents and displaying data from Business Connectivity Services (BCS; a component of SharePoint that allows you to connect to data stored outside SharePoint).

- **Community** – A group of Web Parts for the community features of SharePoint, such as membership, joining a community, and information about the community. In addition, there are tools for community administrators.

- **Content Rollup** – Contains Web Parts that are used to roll up (aggregate) content, such as rolling up search results, providing project summaries, displaying timelines, and showing relevant documents from throughout the site.

- **Document Sets** – Web Parts specifically designed for working with sets of documents.

- **Filters** – Web Parts that can be used to filter information. These Web Parts are designed to be connected with other Web Parts in order to provide a useful filtering mechanism. For example, you might have a list of content and want users to be able to filter based on certain criteria. You could use these Web Parts to provide the filter mechanism.

- **Forms** – Web Parts that allow you to embed HTML or InfoPath forms in a page.

- **Media and Content** – Web Parts that display media, such as images, videos, and pages. In addition, there is also a Web Part for displaying Silverlight applications.

- **PerformancePoint** – Web Parts specifically designed for PerformancePoint services.

- **Project Web App** – Web Parts specifically designed for Project Online. These Web Parts include functionality for displaying information about a project, such as issues, tasks, timesheets, and status.

- **Search** – Provides Web Parts for search functionality, such as the search box for entering a query, search results, and refinement of results.
- **Search-Driven Content** – Provides Web Parts that display content based on search. For example, Web Parts that show items matching a certain tag, pages based on a search query, and recently changed items.
- **Social Collaboration** – Web Parts designed for the social components of SharePoint, such as user contact details, shared note board, tag clouds, and user tasks.

How to Add Web Parts

1. On a **Project Site** page, from the **Settings** menu, click **Edit Page**.
2. On the **Project Site** page, in a **Web Part Zone** (Top, Left or Right), click **Add a Web**.
3. From the **Insert** tab, select a Web Part category, then select the web part and click **Add**.

Adding SharePoint Apps

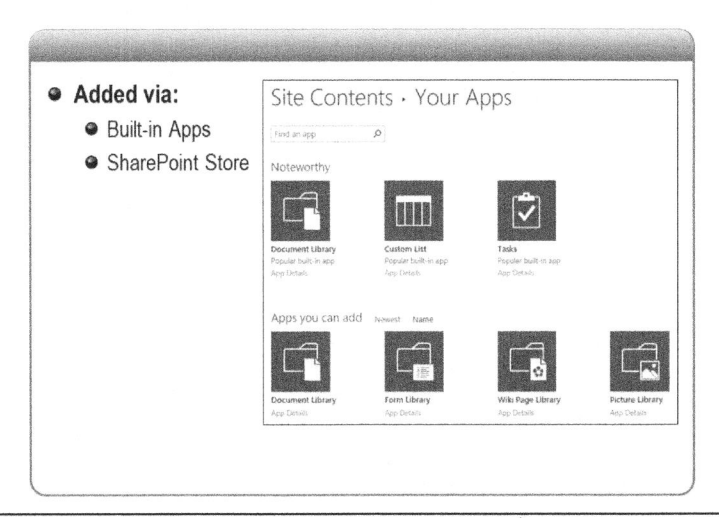

- **Added via:**
  - Built-in Apps
  - SharePoint Store

Apps for SharePoint are small, easy-to-use, stand-alone applications that perform tasks or address specific business needs. You can add apps to your site to customize it with specific functionality or to display information. For example, you can add apps that perform general tasks like time and expense tracking or apps that perform various document-based tasks. You can also add apps that display news or information from third-party websites or that connect to social websites.

You may be able to add apps to your site from a variety of sources. For example, if your organization has developed custom apps for internal business use, you can add these from your organization's App Catalog by browsing the apps under From Your Organization. You can also purchase apps from third-party developers by browsing the SharePoint store. If you've worked with previous versions of SharePoint, note that native SharePoint features such as lists and document libraries are now considered "built-in apps," and you can also add them by using the Add an app command.

How to Add a SharePoint App (built-in)

1. On a **Project Site** page, from the **Settings** menu, click **Add an app**.
2. On the **Site Contents: Your Apps** page, select the app you want to add and complete the app settings.

How to Add a SharePoint App (SharePoint Store)

1. In the Quick Launch, click **SharePoint Store**.
2. On the **SharePoint Store** page, use the **Categories** to filter or you can type the name or tag in the search box.
3. Click the app you want to add.
4. To buy the app, click **Buy It**. If it's a free app, click **Add it**.
5. Follow the steps to log in with your Microsoft account to buy the app.
6. The app will now appear on the **Site Contents** page. You can go to the app by clicking it on the Site Contents Page, which will take you to the app.

Saving the Project Site as SharePoint Template

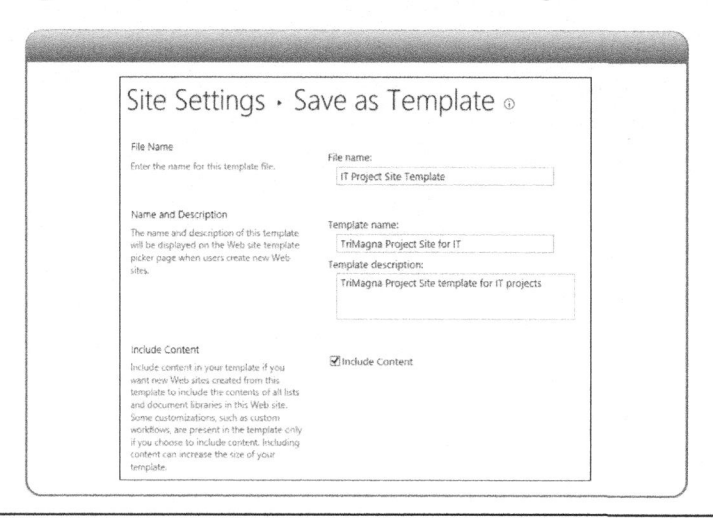

A site template provides the starting point for a new site. Many default site templates are available to suit different needs, and they are available when you create a new site. If you customized your site and want to reuse its settings and structure, you can save it as a site template. If you have permission to customize sites, you can create and apply custom site templates. Site templates are stored and managed in the Site Template Gallery on your site.

How to Save the Project Site as a SharePoint Template

1. On a **Project Site** page, from the **Settings** menu, click **Site settings**.

2. On the **Site Settings** page, under **Site Actions**, click **Save site as template**.

3. On the **Site Settings - Save as Template** page, complete the file name, template name and description and click **OK**.

Practice: Creating a Custom Project Site Template

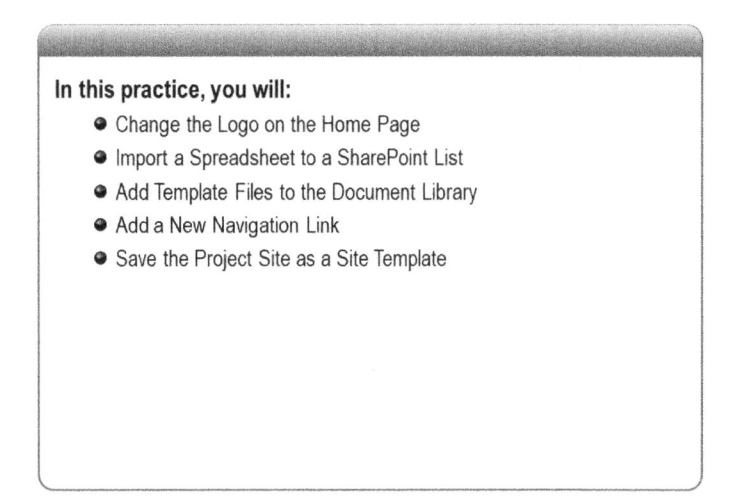

In this practice, you will:
- Change the Logo on the Home Page
- Import a Spreadsheet to a SharePoint List
- Add Template Files to the Document Library
- Add a New Navigation Link
- Save the Project Site as a Site Template

In this practice, you will customize a Project Site by adding/modifying web parts and then save the site as a custom Project Site template.

Exercise 1: Change the Logo on the Home Page

In the exercise, you will change the home page logo.

1. On the **Admin Project** home page, under the **Get started with your site** section, click the **Your site. Your brand** tile.
2. On the **Site Settings: Title, Description, and Logo** page, under **Insert Logo,** click **From Computer**.
3. In the **Add a document** window, click **Browse**.
4. In the **Choose File to Upload** dialog box, navigate to the location where you saved your course files and double click **TriMagna_logo.gif**.
5. In the **Add a document** window, click **OK**.
6. On the **Site Settings: Title, Description, and Logo** page, click **OK**.

Exercise 2: Import a Spreadsheet to a SharePoint List

In the exercise, you will import the contents of an Excel spreadsheet to a SharePoint list.

1. On the **Admin Project** home page, under the **Get started with your site** section, click the **Add list, libraries and other apps** tile.
2. On the **Site Contents: Your Apps** page, in the **Find an app** box, type **Import** and press **Enter**.

3. Click the **Import Spreadsheet** tile.

4. On the **Site Contents: New** page, in the **Name** box, type **TriMagna Contact List** and then click **Browse**.

5. In the **Choose File to Upload** dialog box, navigate to the location where you saved your course files and double click **TriMagna Contact List.xlsx**.

6. On the **Site Contents: New** page, click **Import**.

   This launches Microsoft Excel and opens the TriMagna Contact List.xlsx spreadsheet.

   NOTE: *If you receive the following error message in Internet Explorer, you will need to add the PWA site to your list of Trusted Sites.*

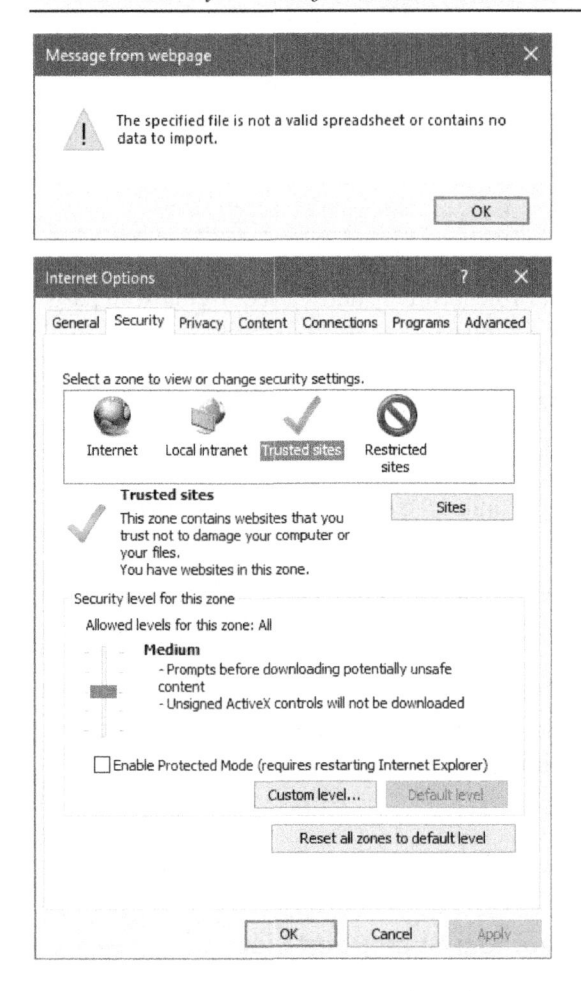

7. In the **Import to Windows SharePoint Services list** dialog box, in the **Range Type** list, select **Range of Cells**, in the **Select Range** box, click the **Range** button (as shown below)

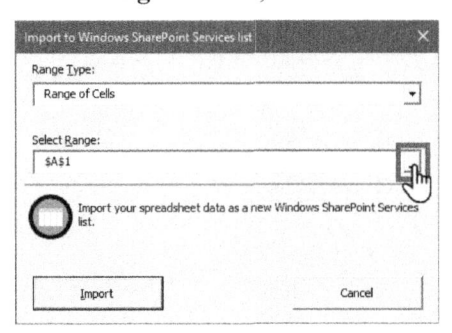

and select the spreadsheet content (A1:F76), then click **Return** button.

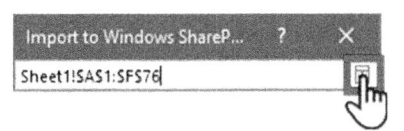

8. In the **Import to Windows SharePoint Services list** dialog box, click **Import**.
9. On the **TriMagna Contact List** page, at the bottom left side of the Quick Launch, click **Return to classic SharePoint**.
10. On the **TriMagna Contact List** page, click the **List** tab and then click **List Settings**.
11. On the **TriMagna Contact List: Settings** page, in the **General Settings** section, click **List name, description and navigation**.
12. On the **List General Settings: TriMagna Contact List** page, in the **Navigation** section, select **Yes** and click **Save**.
13. On the **TriMagna Contact List: Settings** page, click the **TriMagna logo**.

    The TriMagna Contact List is now displayed in the Quick Launch bar.

14. On the **Admin Project** home page, in the **Get started with your site** section, click **REMOVE THIS**.
15. In the **Message from webpage** dialog box, click **OK**.

Exercise 3: Add Template Files to the Document Library

In this exercise, you will add Word template files to the document library.

1. On the **Admin Project** home page, in the quick launch bar, click **Documents**.
2. On the **Documents** page, click the **Files** tab and click **New Folder**.
3. On the **New Folder** page, in the **Name** box, type **Templates** and click **Create**.
4. On the **Documents** page, click **Templates** folder.
5. Launch **File Explorer** and navigate to the location where you saved the course files and select all word documents and then drag and drop the files onto the **Documents: Templates** page, as shown below:

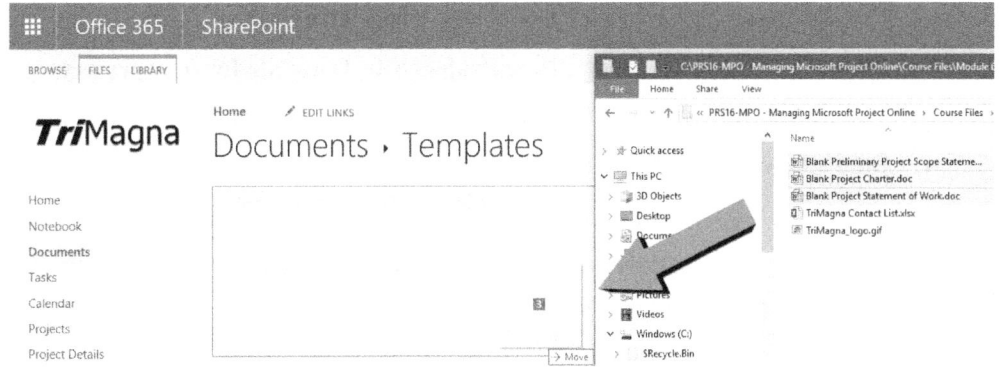

The three Word documents are now listed in the Templates folder.

Exercise 4: Add a New Navigation Link

In this exercise, you will add a new navigation link to the PWA Home page.

1. From the **Templates** page, click **Settings** menu, and then click **Site settings**.

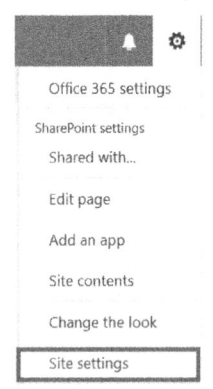

2. On the **Site Settings** page, in the **Look and Feel** section, click **Quick Launch**.

3. On the **Quick Launch** page, in the menu bar, click **New Navigation Link**.

4. On the **New Navigation Link** page, in the **URL** box, type **https://<companyname>.sharepoint.com/sites/pwa**, in the **Description** box, type **Project Web App** and click **OK**.

Exercise 5: Save the Project Site as a Site Template

In this exercise, you will save the TriMagna Admin Project site as a site template.

1. From the **Site Settings – Quick Launch** page, click **Settings** menu, and then click **Site settings**.

2. On the **Site Settings** page, in the **Site Actions** section, click **Save site as template**.

3. On the **Save as Template** page, configure the following settings and click **OK**:

Setting	Perform the following:
File Name	Type **Project Site Template**
Template name	Type **Project Site Template**
Template Description	Type **Project Site Template for new projects**
Include Content	Select **Include Content** check box

4. On the **Operation Completed Successfully** page, click the **solution gallery** link.

   The new project site template is now listed.

5. On the **Solution Gallery** page, click the **Browse** tab and then click the **Project logo** to return to PWA Home page.

Summary

> **In this module, you learned how to:**
> - Work with Project Sites and Elements
> - Create a Custom Project Site Template

In this module, you learned how to work with projects sites and their elements. You also learned how to create a custom project site and then how to create a site template.

Objectives

After completing this module, you learned how to:

- Work with Project Sites and Elements
- Create a Custom Project Site Template

This page is intentionally left blank

Module 7: Project Online Administration

Contents

Module Overview 1
Lesson 1: Working with Project Online Workflows 2
    Overview of Workflows in Project Online 3
    Modifying Workflow Phases 4
    Creating Workflow Stages 5
    Creating Project Detail Pages 6
    Creating Enterprise Project Types 7
    Creating a Workflow 9
    Troubleshooting Project Online Workflows 10
    Change or Restart Workflows 12
    Practice: Customizing Workflows and PDPs 13
Lesson 2: Sharing Project Online with External Users 22
    Overview of Project / SharePoint Online B2B Sharing 23
    Configuring Sharing on PWA Site Collection 24
    Configuring Sharing on Project Web App 26
    Adding External Users in PWA 28
    Assigning Project Online Licensing for External Users 29
    Practice: Sharing Project Online with External Users 30
Lesson 3: Managing Queue Jobs and Enterprise Objects 33
    Managing Queue Jobs 34
    Deleting Enterprise Objects 35
    Forcing Check-In of Enterprise Objects 36
    Practice: Working with Enterprise Objects 37
Lesson 4: Troubleshooting Resources 39
    Office 365 Service Health 40
    Office 365 Message Center 41
    Office 365 Usage Reports 42
    Tune Project Online Performance 43
Summary 44

EXCLUSIVELY PUBLISHED BY

PMO Logistics
679 Roberta Avenue
Winnipeg, Manitoba, Canada R2K 0K9

Copyright © 2017 by Roland Perreaux

All rights reserved. No part of the contents of this document may be reproduced or transmitted in any form or by any means without written permission of the publisher.

Information in this document, including URL and other Internet Web site references, is subject to change without notice. Unless otherwise noted, the example companies, organizations, products, domain names, e-mail addresses, logos, people, places, and events depicted herein are fictitious, and no association with any real company, organization, product, domain name, e-mail address, logo, person, place, or event is intended or should be inferred. Complying with all applicable copyright laws is the responsibility of the user. Without limiting the rights under copyright, no part of this document may be reproduced, stored in or introduced into a retrieval system, or transmitted in any form or by any means (electronic, mechanical, photocopying, recording, or otherwise), or for any purpose, without the express written permission of PMO Logistics Inc.

The names of manufacturers, products, or URLs are provided for informational purposes only and PMO Logistics makes no representations and warranties, either expressed, implied, or statutory, regarding these manufacturers or the use of the products with any Microsoft technologies. The inclusion of a manufacturer or product does not imply endorsement of Microsoft of the manufacturer or product. Links are provided to third party sites. Such sites are not under the control of PMO Logistics and PMO Logistics is not responsible for the contents of any linked site or any link contained in a linked site, or any changes or updates to such sites. PMO Logistics is not responsible for webcasting or any other form of transmission received from any linked site. PMO Logistics is providing these links to you only as a convenience, and the inclusion of any link does not imply endorsement of PMO Logistics of the site or the products contained therein.

PMO Logistics may have patents, patent applications, trademarks, copyrights, or other intellectual property rights covering subject matter in this document. Except as expressly provided in any written license agreement from PMO Logistics, the furnishing of this document does not give you any license to these patents, trademarks, copyrights, or other intellectual property.

PMO Logistics, Professional Training Series, Upgrading Skills Series, TriMagna Corporation and TriMagna Corporation logo are either registered trademarks or trademarks of PMO Logistics Inc. in Canada, the United States and/or other countries.

Microsoft, Active Directory, Internet Explorer, Outlook, Project Server, SharePoint, SQL Server, Visual Studio, Windows and Windows Server are either registered trademarks or trademarks of Microsoft Corporation in the United States and/or other countries.

All other trademarks are property of their respective owners.

Author: Rolly Perreaux, PMP, MCSE, MCT

Publisher: PMO Logistics
Developmental Editor: Heather Perreaux
Cover Graphic Design: Andrea Ardiles
Technical Testing: Underground Studioworks

Post-Publication:
Errata List Contributors:

Module Overview

- Working with Workflows and PDPs
- Sharing Project Online with External Users
- Managing Queue Jobs and Enterprise Objects
- Troubleshooting Resources

In this module, you will work with Project Online Workflows and all the corresponding elements. You will then learn how to share Project Online with external users of your organization. You will also learn how to manage queue jobs and enterprise objects and the troubleshooting resources available for a PWA Administrator.

Objectives

After completing this module, you will:

- Work with Workflows and PDPs
- Share Project Online with External Users
- Manage Project Server Queue Jobs and Enterprise Objects
- Learn Troubleshooting Resources

Lesson 1: Working with Project Online Workflows

- Overview of Workflows in Project Online
- Modifying Workflow Phases
- Creating Workflow Stages
- Creating Project Detail Pages
- Creating Enterprise Project Types
- Creating Workflows
- Troubleshooting Project Online Workflows
- Change or Restart Workflows

In this lesson, you will learn how Workflows work on Project Online. You will learn how to work with Workflow Stages and Phases. You will learn how to create Project Detail Pages (PDP) and Enterprise Project Types. You will also learn how to create a workflow using all these elements together and how to troubleshoot Project Online workflows.

Objectives

After completing this lesson, you will be able to:

- Know how Workflows work in Project Online
- Modify Workflow Phases
- Create Workflow Stages
- Create Project Detail Pages
- Create Enterprise Project Types
- Create a Workflow
- Troubleshoot Project Online Workflows
- Change or Restart Workflows

Overview of Workflows in Project Online

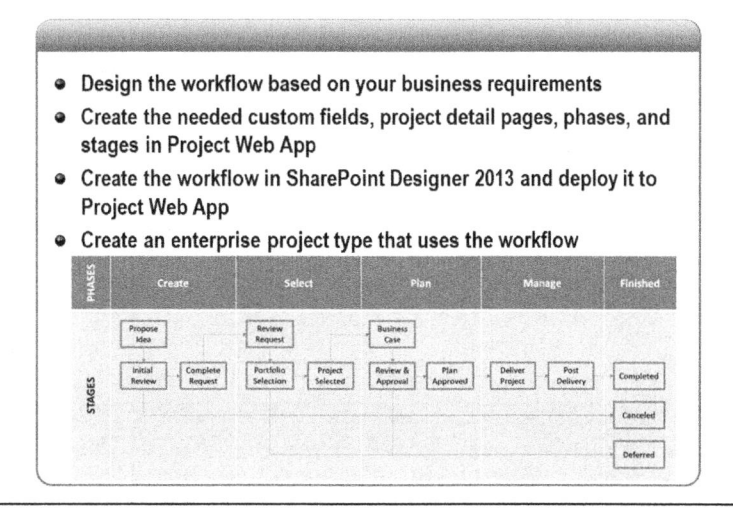

- Design the workflow based on your business requirements
- Create the needed custom fields, project detail pages, phases, and stages in Project Web App
- Create the workflow in SharePoint Designer 2013 and deploy it to Project Web App
- Create an enterprise project type that uses the workflow

Workflows enforce your business processes and provide a structured way for projects to move through phases and stages. You can set up a workflow to perform a variety of actions based on the user input for each stage, including sending emails, assigning tasks, and waiting for specific project actions.

For example, you might have an "Initial Proposal" stage where you include a project detail page with a custom field for project cost. You could configure the workflow to automatically accept or reject the project based on whether the project cost exceeds a certain limit.

Below are the five phases of demand management that are included in Project Web App and how they fit together. Within each phase are example stages such as "Propose idea" and "Initial review." Each stage can have one or more associated project detail pages. The entire collection of stages represents a single workflow that can be associated with an EPT.

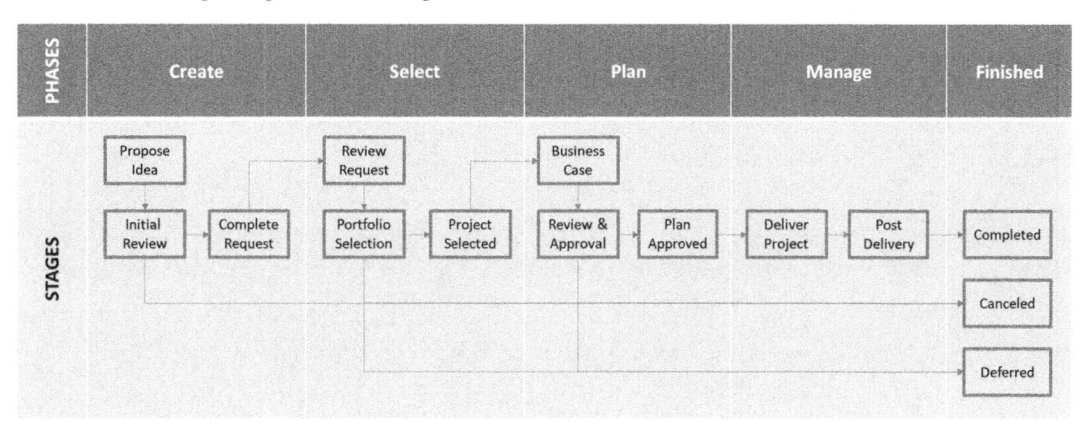

There are four general steps to create your workflow in Project Web App:

- Design the workflow based on your business requirements.
- Create the needed custom fields, project detail pages, phases, and stages in Project Web App.
- Create the workflow in SharePoint Designer 2013 and deploy it to Project Web App.
- Create an enterprise project type that uses the workflow.

Modifying Workflow Phases

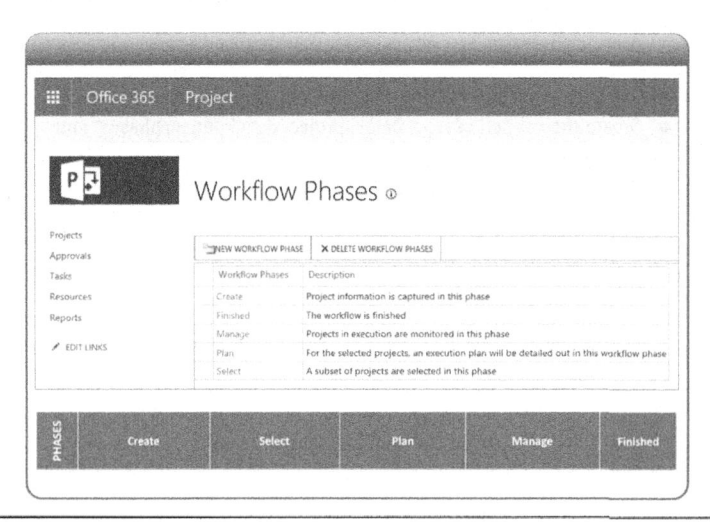

A Workflow Phase is used to organize multiple stages that make up a common set of activities in the project life cycle. Examples of phases are project creation, project selection, project plan, project management and project finish (represented in the default Project Web App phases as Create, Select, Plan, Manage and Finished).

The phases themselves are just a way of organizing your stages and do not determine the order in which the stages are executed. (The order of the stages is determined by the associated workflow.)

Creating Workflow Stages

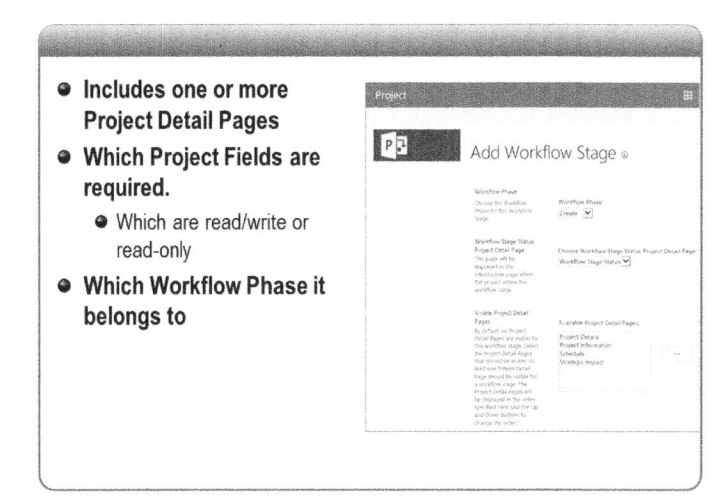

- **Includes one or more Project Detail Pages**
- **Which Project Fields are required.**
  - Which are read/write or read-only
- **Which Workflow Phase it belongs to**

A Stage includes one or more project detail pages, grouped to gather information about a project. This information can be used or updated by a workflow.

In Project Web App, you can define which project detail pages are displayed in each stage, which fields are required, and which are read/write or read-only, and which phase the stage is part of.

For each stage of a project, you need to define what actions need to occur and based on the business requirements of the project, what information needs to be gathered. This information will help you define the list of fields that you need to display in the project detail pages and what actions you need the workflow perform. The business requirements for a stage can be defined as:

- What needs to happen with the project in each stage
- What is required information that you want to capture using project detail pages
- What is the state of the fields in each stage (Required, read/write, or read-only)

Creating Project Detail Pages

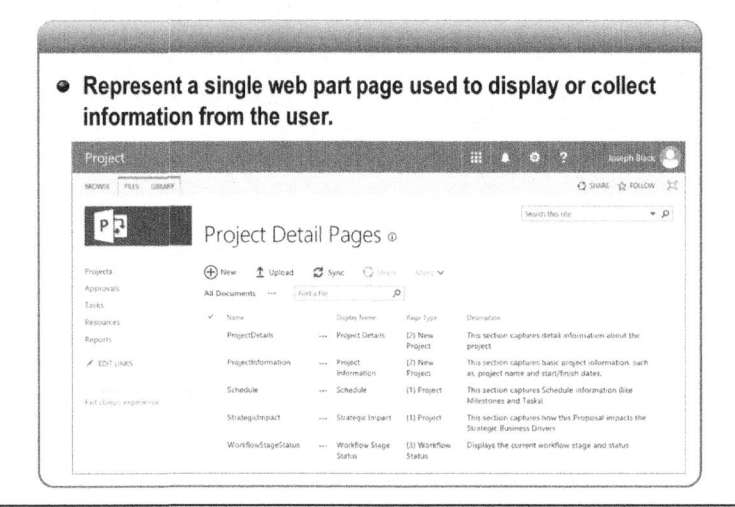

- **Represent a single web part page used to display or collect information from the user.**

Project Detail Pages (PDPs) represent a single web part page used to display or collect information from the user. You can create PDPs much as you create any Web Part Page in a SharePoint site, where you can add Web Parts that provide the experience that you want. You can add individual Web Parts from the Web Part galleries or create custom Web Parts.

There are three types of Project Detail Pages that can be created:

- **New Project** – Used for creating a project. This type of PDP is required with an enterprise project template that has a workflow for portfolio analysis.
- **Workflow Status** – Shows the current stage and status for a project proposal.
- **Project** – Used for editing.

To Create a Project Detail Page:

1. On the PWA home page, click **Settings → PWA Settings**.
2. On the **Server Settings** page, in the **Workflow and Project Detail Pages** section, click **Project Detail Pages**.
3. Click the **Documents** tab.
4. Select **New Document** on the ribbon.
5. On the newly created blank page, select **Add a Web Part**.
6. In **Categories**, select **Project Web App**, and then select the Web Part you want to add.
7. From the upper-right menu, select **Edit Web Part**.
8. In the **Edit Web Part** pane, click the **Modify** button.
9. Select the project data that you want to display, and then click **OK**.
10. In the **Edit Web Part** pane, under **Appearance**, enter a title and then click **OK**.
11. Select the **Page Tools** tab on the ribbon and then click **Stop Editing**.
12. Now select **Edit Properties** on the ribbon and update the **Display Name** and **Page Type** for your newly created **Project Detail Page**.

Creating Enterprise Project Types

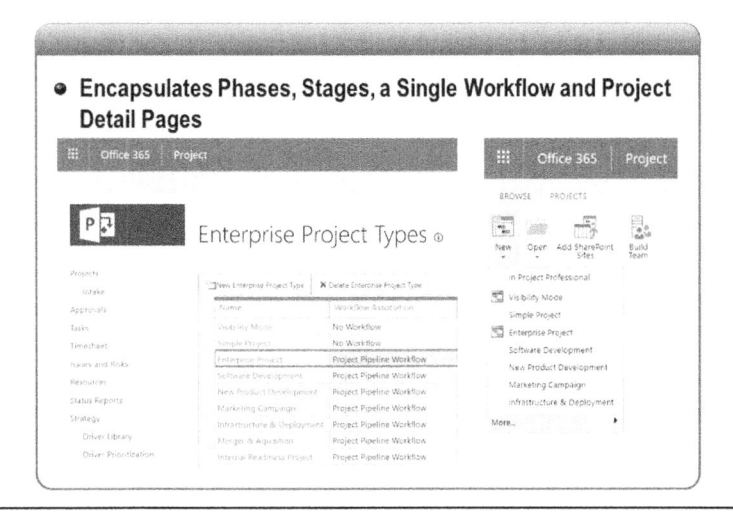

- Encapsulates Phases, Stages, a Single Workflow and Project Detail Pages

An enterprise project type (EPT) represents a wrapper that encapsulates phases, stages, a single workflow, and Project Detail Pages (PDPs). Each EPT represents a single project type. Normally, project types are aligned with individual departments: for example, marketing projects, IT projects, or HR projects. Using project types helps categorize projects within the same organization that have a similar project life cycle. For a user, the EPTs appear in a drop-down list of project types when the user clicks New Project on the ribbon in PWA.

To Create a New Enterprise Project Type

1. On the PWA home page, click **Settings → PWA Settings**.

2. On the **Server Settings** page, in the **Workflow and Project Detail Pages** section, click **Enterprise Project Types**.

3. Click **New Enterprise Project Type**. In the Name box, enter a name for the type, and then provide a brief description in the Description box.

4. If you want to generate a Project ID for each project created by this EPT, do the following in the Project ID section:

   NOTE: Project IDs are not required to create your EPT.

   a. In the **Prefix** field, enter characters that will be at the start of each generated Project ID.

   b. In the **Starting Number** field, enter a number that will serve as a starting point for Project IDs that are going to be generated for this EPT.

   c. In the **Postfix** field, you can enter characters that can be used to append your Project IDs that are generated by this EPT.

   d. In the **Minimum Digit Padding** field, enter a number that will be used to increment the Starting Number for newly generated Project IDs.

5. Select a workflow from the **Site Workflow Association** list. Once you associate a site workflow with a project type and save the type, you cannot go back and update the type to use a different workflow.

6. Select a project detail page from the **New Project Page** list. This is the first page that users will see when they create a new project in the Project Center through the EPT.

7. Click the button next to the **Departments** field to select the departments that you want to associate with this project type, if it is appropriate.

8. If you want to associate an image with this project type to display in the new project menu, provide the URL for the image in the **Type the URL** box.

9. In the **Order** section, choose whether you want this project type to appear at the end of the list of project types, or if you want to control its placement in the list.

10. In the **Site Creation** section, select when a project site is created for your new projects. You can choose to:

    - Automatically create a site when the project is first published.
    - Allow the user to create a project site for their project on publish.
    - Disallow the user to create a project site for their project.

11. In the **Site Creation Location** section, you can specify the location in which your project sites will be created in the Location URL field. Project sites will be created as subsites in the specified location.

    NOTE: If you decide to create your project sites outside the PWA site collection, you will not be able to synchronize your user permissions or SharePoint Tasks Lists.

    NOTE: The **Site Creation Location** section will visible only in Project permissions mode.

12. In the **Synchronization** section:

    - Select **Sync User Permissions** to grant project resources access to the project site.
    - Select **Sync SharePoint Tasks Lists** to copy tasks to the SharePoint Tasks List in your project site. Only project sites that are created inside of the Project Web App site will synchronize.

13. In the **Site Language** section, select the default language in which your sites will be provisioned from the **Default site template language** list.

14. If appropriated, select a template from the **Project Plan Template**.

15. Click **Save** to save this enterprise project type to the server, thus making it available for new project or proposal creation.

Creating a Workflow

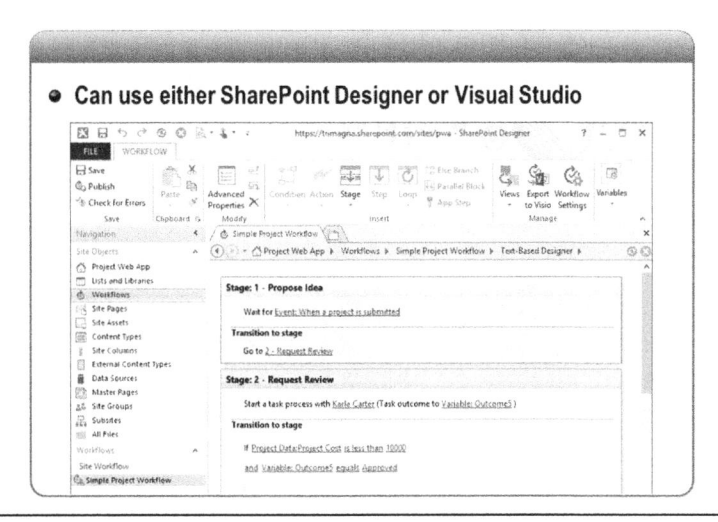

- Can use either SharePoint Designer or Visual Studio

Workflows enable organizations to reduce the amount of unnecessary interactions between people as they perform business processes. For example, to reach a decision, groups typically follow a series of steps. The steps can be a formal, standard operating procedure, or an informal implicitly understood way to operate. Collectively, the steps represent a business process.

To Create a Project Online Workflow with SharePoint Designer

1. In Project Web App, create the following elements that a workflow requires:

    a. Review the existing workflow phases; create phases as necessary.

    b. Create the enterprise custom fields that the workflow will use. To be available in a workflow stage, a custom field must be controlled by a workflow.

    c. Edit or create the project detail pages (PDPs) that your workflow stages will use to collect information for the project.

    d. Create the necessary workflow stages, and then associate each workflow stage with the correct phase.

2. Launch SharePoint Designer 2013:

    a. Open the Project Web App site, and then create a site workflow that uses the SharePoint 2013 Workflow - Project Server workflow platform.

    b. Add the stages that the workflow uses.

    c. Insert the workflow steps, conditions, actions, and loops that are required in each stage.

    d. Check for any workflow errors and fix any that you find, then Publish the workflow.

    e. After it is published, the workflow shows in the list of workflows for the Project Web App site.

Troubleshooting Project Online Workflows

- Create a Project Center View to see Workflow State
- Reviewing the Errors
- Acting on the Errors and Taking Additional Steps

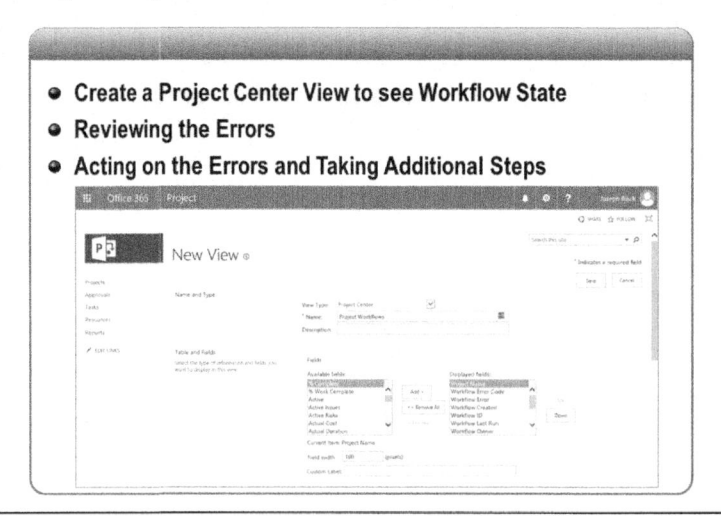

NOTE: *Excerpt from Microsoft Office Support website article, titled: Troubleshooting Project Online workflows, November 30, 2017, https://support.office.com/en-us/article/Troubleshoot-Project-Online-workflows-fd893a1f-26c3-48c0-a854-4aa12fd808b1.*

It is common as a PWA administrator to have to troubleshoot Project workflows. Depending on how the workflow process has been defined for the organization, there may be instances where an admin needs to act for the workflow instance to progress. There are fields that a user can add to a Project Center view to better understand the state of each Project workflow. With the information provided in these fields, a Project Web App admin can take the appropriate corrective action to unblock the progression of the Project workflow.

There are three steps on troubleshooting Project Online workflows:

- Create a Project Center View to see Workflow State
- Reviewing the Errors
- Acting on the Errors and Taking Additional Steps

Create a Project Center View to See Workflow State

1. On the PWA Home page, click **Settings → PWA Settings**.
2. On the **PWA Settings** page, under the **Look and Feel** section, click **Manage Views**.
3. On the **Manage Views** page, click **New View**.
4. On the **New View** page, in the **View Type** list, select **Project Center**.
5. In the **Name** box, type **Project Workflows**.
6. In the **Table and Fields** section, from the **Displayed fields** list, remove the **Start** and **Finish** fields.
7. Add the following fields from the **Available fields** list:

Workflow Error Code	Workflow Phase Name
Workflow Error	Workflow Stage Name

Module 7: Project Online Administration

Workflow Created	Workflow State
Workflow ID	Checked Out
Workflow Last Run	Checked Out By
Workflow Owner	

8. Scroll to the bottom of the page and click **Filter**.

9. In the **Custom Filter** window, in the **Field Name** list, select **Workflow Error Code**, in the **Test** list, select **is greater than**, in the **Value** box, type **1** and click **OK**, as shown below:

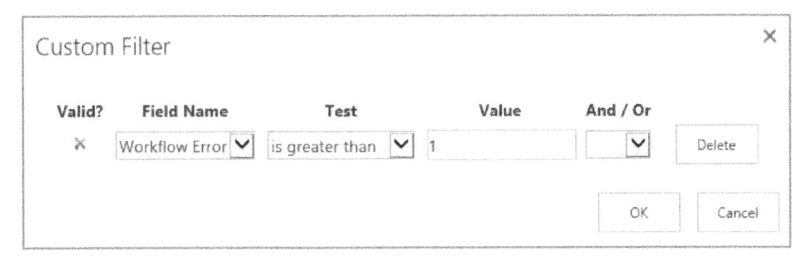

```
Custom Filter                                              ×

Valid?   Field Name        Test            Value    And / Or
  ✗      Workflow Error ▼  is greater than ▼  1       ▼      Delete

                                              OK      Cancel
```

10. In the **Security Categories** section, in the **Available Categories** list, select **My Organization** and click **Add >**.

11. Click **Save**.

> *NOTE: You will receive the following message:*
> *"You have not assigned a security category to this view. Failure to do so will prevent anyone from seeing the view in the dropdown or from using the view. Do you want to save anyway?"*

Click **OK**, because only members of the PWA Administrators group will be only be able to view the Project Workflows Project Center view.

Reviewing Workflow Errors

If the admin created a Project Center view as described above, the view can be accessed following these steps:

1. On the PWA Home page, in the quick launch, click **Projects**.

2. Click on the **Projects** tab in the ribbon.

3. Select the **Project Workflows** view

This will provide the PWA Administrator with a list of all the projects and the current status of each project's workflow, including errors.

Acting on the Errors and Taking Additional Steps

The complete list of errors and subsequent actions to be taken, can be viewed at:

https://support.office.com/en-us/article/Troubleshoot-Project-Online-workflows-fd893a1f-26c3-48c0-a854-4aa12fd808b1#actingonerrors

Change or Restart Workflows

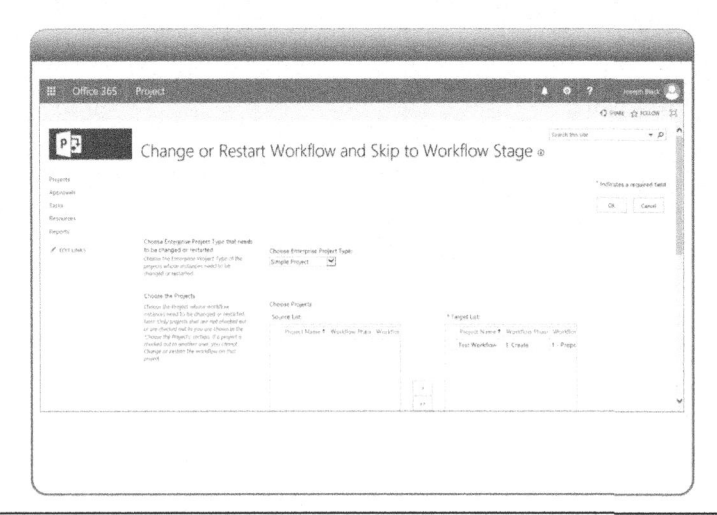

There are times when you need to restart a workflow. Restarting a workflow will cause the workflow engine to execute the workflow from the very beginning. The good news when you restart a workflow is that the project data previously entered is not lost or reset. This action simply tells the workflow to "Go to Stage 1" and execute everything again. Similarly, changing a project to another Enterprise Project Type that has a workflow, will also cause the project to execute the new workflow from the very beginning.

To Change or Restart a Workflow

4. On the **PWA home** page, click **Settings → PWA Settings**.

5. On the **PWA Settings** page, in the **Workflow and Project Detail Pages** section, click **Change or Restart a Workflow**.

6. Under **Choose Enterprise Project Type** list, select the Enterprise Project Type that needs to be changed or restarted.

7. Under **Choose Projects** list, select the project with the workflow instances that need to be changed or restarted, and add it to the **Target List**.

   NOTE: Only projects that are not checked out or are checked out to you are shown in the Choose Projects section. If a project is checked out to another user, the workflow cannot be changed or restarted on that project.

8. Under **Choose new Enterprise Project Type or restart workflow**, select whether to restart the current workflow for the selected projects or to associate the projects with a new Enterprise Project Type.

   If a new Enterprise Project Type is selected, you must then select from the drop-down menu the new Enterprise Project Type that you wish to use.

9. Under **Choose Workflow Stage**, select the target workflow stage that this workflow will skip to.

10. Click **OK**.

Practice: Customizing Workflows and PDPs

In this practice, you will:
- Modify Workflow Phases
- Create Workflow Stages
- Install SharePoint Designer 2013
- Publish an Empty Workflow to Project Web App
- Create an Enterprise Project Type
- Create a Custom Field and Add to a PDP
- Add Stages, Actions and Conditions to a Workflow
- Test the Workflow

In this practice, you will modify the built-in Phases and create new stages. You will then modify a built-in PDP, install SharePoint Designer 2013 and create a new enterprise project type. You will then use SharePoint Designer 2013 to create a custom project workflow and then test the functionality of the project workflow.

Exercise 1: Modify Workflow Phases

In this exercise, you will modify the names of the Workflow Phases.

1. In the PWA site, click **Settings → PWA Settings**.
2. On the **PWA Settings** page, under **Workflow and Project Detail Pages**, click **Workflow Phases**.
3. Name the Workflow Phases to the following:

Workflow Phases	Renamed to
Create	**1. Create**
Finished	**5. Finished**
Manage	**4. Manage**
Plan	**3. Plan**
Select	**2. Select**

Exercise 2: Create Workflow Stages

In this exercise, you will create Workflow Stages and assign them to a Workflow Phase.

1. In Project Web App, click **Settings → PWA Settings**.
2. Under **Workflow and Project Detail Pages**, click **Workflow Stages**.
3. Click **New Workflow Stage**.
4. On the **Add Workflow Stage** page, complete the form with the following settings:

Setting	Perform the following:
Name	Type **1 – Propose Idea**
Workflow Phase	Select **1. Create**
Available Project Detail Pages	Select **Project Information** and click **>**.

5. Leave the other options at their default values and click **Save**.
6. Repeat Steps 3-5 with the following settings:

Stage Name	Workflow Phase	Available Project Detail Pages
2 – Request Review	2. Select	Project Information
3 – Plan	3. Plan	Schedule Project Information
4 – Execute	4. Manage	Schedule Project Information
5 – Cancelled	5. Finished	Project Information
6 – Complete	5. Finished	Schedule Project Information

Exercise 3: Install SharePoint Designer 2013

In this exercise, you will install SharePoint Designer 2013 from the Office.com website.

1. Open a new tab in your web browser and go to www.office.com and sign in as the PWA Administrator.
2. On the Office home page, click **Install Office** apps and click **Other install options** OR click **Other installs**, as shown below:

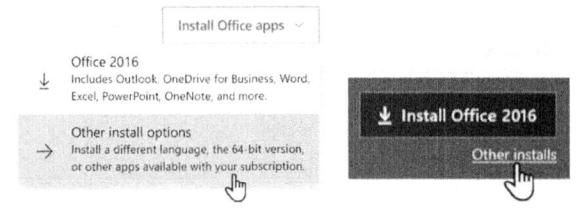

3. On the **Software** page, click **Tools & add-ins** and under **SharePoint Designer 2013**, click **Download and install**.
4. Follow the web instructions to install SharePoint Designer 2013.

Exercise 4: Publish an Empty Workflow to Project Web App

In this exercise, you will create an empty workflow to be used to link to an enterprise project type that will be created in the next exercise.

1. Start **SharePoint Designer 2013**.
2. Click **Open Site**.
3. In the **Open Site** dialog box, type the URL of your PWA site and then click **Open**.

4. In the navigation pane, click **Workflows**.
5. On the ribbon, click **Site Workflow**.
6. In the **Create Site Workflow** window, in the **Name** box, **Simple Project Workflow** and in the **Platform Type** list, ensure that **SharePoint 2013 Workflow - Project Server** is selected and then click **OK**.
7. On the ribbon, click **Publish**.
8. Close SharePoint Designer.

Exercise 5: Create an Enterprise Project Type

In this exercise, you will create an Enterprise Project Type and use the new Site Workflow created in the previous exercise.

1. Switch tab to PWA home, click **Settings → PWA Settings**.
2. On the **PWA Settings** page, under the **Workflow and Project Detail Pages** section, click **Enterprise Project Types**.
3. Click **New Enterprise Project Type**.
4. On the Add Enterprise **Project Type** page, complete the form with the following settings:

Setting	Perform the following:
Name	Type **Simple Project**
Description	Type **This project type is used for all projects which has an estimated cost of less than $10,000. The Director of Programs will be the approver.**
Site Workflow Association	Select **Simple Project Workflow**
New Project Page	Select **Project Information**
Site Creation	Select **Do not create a site**.

5. Click **Save**.

Exercise 6: Create a Custom Field and Add to a PDP

In this exercise, you will create a custom project field called Project Cost and then add the custom field to the Project Information PDP.

1. Click **Settings → PWA Settings**.
2. On the **PWA Settings** page, under the **Enterprise Data** section, click **Custom Fields and Lookup Tables**.
3. Under **Enterprise Custom Fields**, click **New Field**.
4. In the **Name** box, type **Project Cost**.
5. In the **Description** box, type **Estimated cost of the project.**
6. Under **Entity and Type**, in the **Type** list, select **Cost**.
7. Under **Behavior**, select the check box **Behavior controlled by workflow**.
8. Click **Save**.

9. Click **Settings → PWA Settings**.

10. On the **PWA Settings** page, under **Workflow and Project Detail Pages** section, click **Project Detail Pages**.

11. On the **Project Detail Pages** page, click **ProjectInformation**.

12. From the **Page** tab, click **Edit Page**, as shown below:

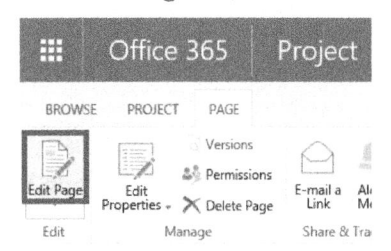

13. In the upper right corner of the **Basic Info** area, open the dropdown menu and choose **Edit Web Part**.

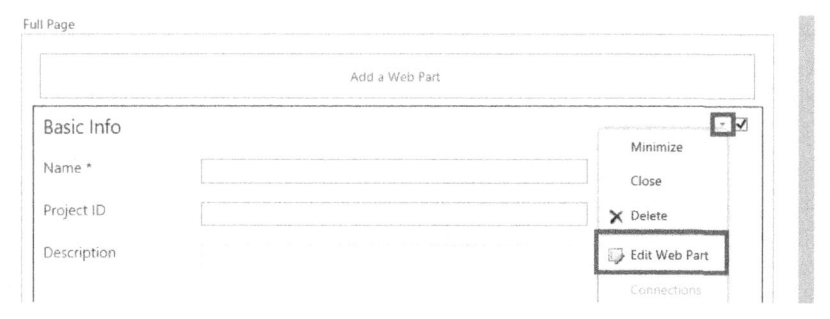

14. Under **Displayed Project Fields**, click **Modify**.

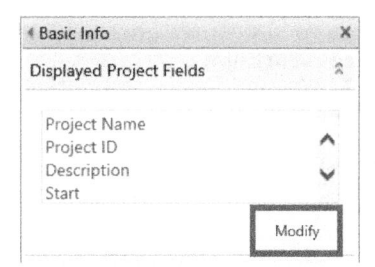

15. In the **Choose Project Fields** list, find the **Project Cost** field and then click the add (>) button to add it to the **Selected Project Fields** list and click **OK**.

16. Scroll down the **Basic Info** web part and click **OK**.

17. On the ribbon, click **Stop Editing**.

Exercise 7: Assign an Approval Task in the Workflow

In this exercise, you will edit the empty workflow you created at the beginning of the practice. You will then add stages, actions and conditions to the site workflow.

1. Start **SharePoint Designer 2013**.

2. On your list of **Recent Sites**, click on your Project Web App site.

3. Sign in as the PWA Administrator.

4. In the **Navigation** pane, click **Workflows**.

Module 7: Project Online Administration   7-17

5. In the details pane, click **Simple Project Workflow**.

6. In the ribbon, click **Edit Workflow**.

7. On the ribbon, click **Stage**, and then click **1 – Propose Idea**.

8. Place the orange cursor in the top section of **Stage 1**, then from the ribbon, click **Action**, then under **Project Web App Actions**, click **Wait for Project Event**, as shown below:

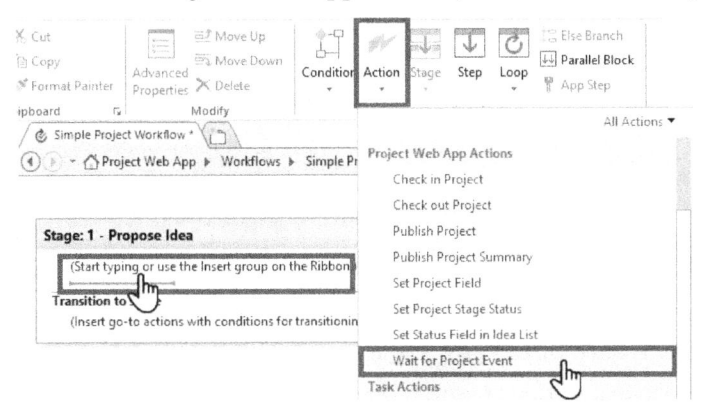

9. Click **this project event**, and select **Event: When a project is submitted**.

10. Click below **Stage 1**, and from the ribbon, click **Stage** and then click **2 – Request Idea**.

11. Click below **Stage 2**, and from the ribbon, click **Stage** and then click **3 – Plan**.

12. Click below **Stage 3**, and from the ribbon, click **Stage** and then click **4 – Execute**.

13. Click below **Stage 4**, and from the ribbon, click **Stage** and then click **5 – Cancelled**.

14. Click below **Stage 5**, and from the ribbon, click **Stage** and then click **6 – Finished**, as shown below:

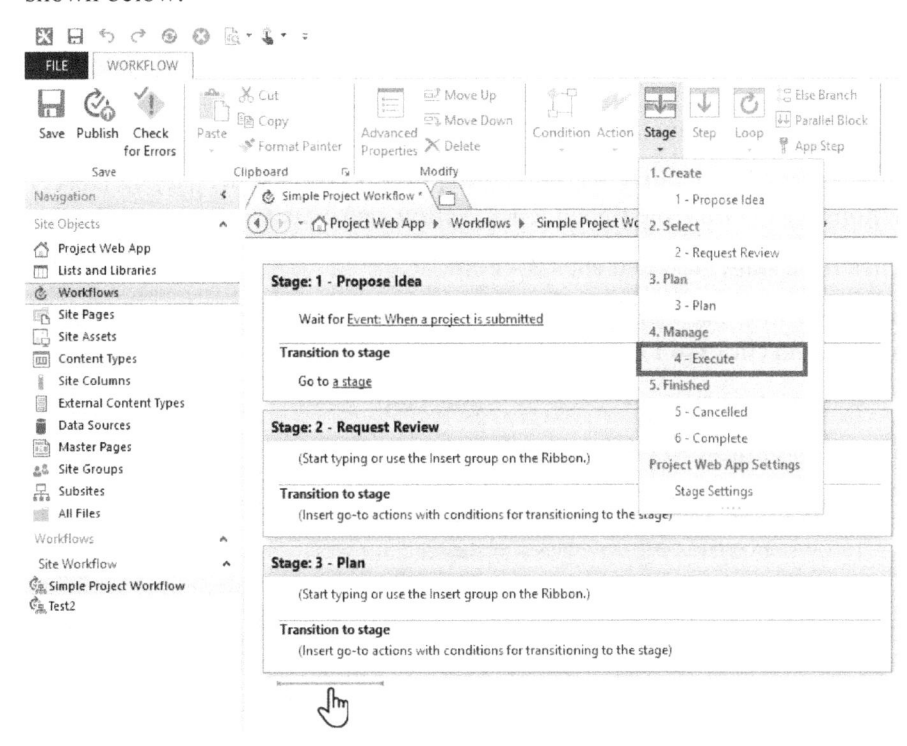

15. In **Stage 1 – Propose Idea**, click in the **Transition to stage** area, then from the ribbon, click **Action** and then click **Go to a stage**.

16. Click the **a stage** link and then click **2 – Request Review**.

Stage 2 – Request Review

17. Click in the top section of **Stage 2 – Request Review**, then from the ribbon, click **Action**, then under **Task Actions**, click **Start a task process**.
18. Click the **these users** link.
19. In the **Start a Task Process** dialog box, click the ellipsis (...) for **Participants** and select **Karla Carter** and click the **Add > >** button and click **OK**.
20. For **Task Title**, click **fx**,
21. In the **Lookup for String** dialog box, in the **Data source** field, select **Project Data**, in the **Field from source** field, select **Project Name** and click **OK**.
22. In the **Start a Task Process** dialog box, click **OK**.
23. Click in the **Transition to stage** section of **Stage 2 – Request Review**, then from the ribbon, click **Condition** and click **If any value equals value**.
24. Click the first **value** link, and then click **fx**.
25. In the **Define Workflow Lookup** dialog box, in the **Data source** field, select **Project Data**, in the **Field from source** field, select **Project Cost** and click **OK**.
26. Click **equals** and click **is less than**
27. Click the second **value** link and type **10000**.
28. From the ribbon, click **Condition** and click **If any value equals value**.
29. Click the first **value** link, and then click **fx**.
30. In the **Define Workflow Lookup** dialog box, in the **Data source** field, select **Workflow Variables and Parameters**, in the **Field from source** field, select **Variable: Outcome** and click **OK**.
31. Click the second **value** link and click **Approved**.
32. After the condition **and Variable: Outcome equals Approved**, click **(Insert go-to actions…)** and from the ribbon, click **Action**, and click **Go to a stage**.
33. Click the **a stage** link, and click **3 - Plan**.
34. Below the Else statement, click **(Insert go-to actions…)** and from the ribbon, click **Action**, and click **Go to a stage**.
35. Click the **a stage** link, and click **5 - Cancelled**.

Stage 2 – Request Review should look like this:

```
Stage: 2 - Request Review
    Start a task process with Karla Carter (Task outcome to Variable: Outcome )
Transition to stage
    If Project Data:Project Cost is less than 10000
    and Variable: Outcome equals Approved
        Go to 3 - Plan
    Else
        Go to 5 - Cancelled
```

Module 7: Project Online Administration

Stage 3 – Plan

36. Place the orange cursor in the top section of **Stage 3 – Plan**, then from the ribbon, click **Action**, then under **Project Web App Actions**, click **Wait for Project Event**.

37. Click **this project event**, and select **Event: When a project is submitted**.

38. Click in the **Transition to stage** section of **Stage 3 – Plan**, then from the ribbon, click **Action** and click **Go to stage**.

39. Click **a stage**, and then click **4 – Execute**.

Stage 4 – Execute

40. Place the orange cursor in the top section of **Stage 4 – Execute**, then from the ribbon, click **Action**, then under **Project Web App Actions**, click **Wait for Project Event**.

41. Click **this project event**, and select **Event: When a project is submitted**.

42. Click in the **Transition to stage** section of **Stage 4 – Execute**, then from the ribbon, click **Action** and click **Go to stage**.

43. Click **a stage**, and then click **6 - Complete**.

Stage 5 – Cancelled

44. Click in the **Transition to stage** section of **Stage 5 – Cancelled**, then from the ribbon, click **Action** and click **Go to stage**.

45. Click **a stage**, and then click **End of Workflow**.

Stage 6 – Complete

46. Place the orange cursor in the top section of **Stage 6 – Complete**, then from the ribbon, click **Action**, then under **Project Web App Actions**, click **Wait for Project Event**.

47. Click **this project event**, and select **Event: When a project is submitted**.

48. Click in the **Transition to stage** section of **Stage 6 – Cancelled**, then from the ribbon, click **Action** and click **Go to stage**.

49. Click **a stage**, and then click **End of Workflow**.

50. From the ribbon, click **Check for Errors**.

51. If the workflow has errors, fix them.
    If the workflow contains no errors, then click **Publish**,

52. Close **SharePoint Designer**.

Exercise 7: Testing the Simple Project Workflow

In this exercise, you will test Simple Project Workflow to ensure that all the elements work properly.

1. Start a Private Mode web browser session and go to the PWA site.

2. Sign in as **BQuinlan@<companyname>.onmicrosoft.com**

3. From the **PWA Home** page, in the quick launch, click **Projects**.

4. From the **Projects** tab, click the **Project** tab and click **New**, and click **Simple Project Type**.

5. On the **Create a new project** page, in the **Name** box, type **Test Workflow** and in the **Project Cost** box, type **9999** and click **Finish**.

6. On the **Workflow Status** page, expand **All Workflow Stages** at the bottom of the page.

   Notice the **State** column displays **In Progress (Waiting for input),** that is because the proposal has not been submitted yet.

7. From the **Project** tab, in the **Workflow** section, click **Submit**.

8. In the **Message from webpage** dialog box, click **OK**.

9. On the **Workflow Status** page, expand **All Workflow Stages** at the bottom of the page.

   Notice that the **State** column in the **1 – Propose Idea** stage now displays **Completed** and now **2 - Request Review** displays **In Progress (Waiting for input)**.

   Now Karla Carter must review the project and workflow.

10. Sign out as **Beth Quinlan (BQuinlan)** from PWA and sign back in to PWA as **Karla Carter (KCarter)**.

11. From the **PWA Home** page, in the quick launch, click **Approvals**.

12. In the **Approvals** page, from the **Approvals** tab, click **Workflow Approvals**.

13. In the **Project Server Workflow Tasks** page, in the menu bar, click **All Tasks**.

14. Select **Test Workflow**, and click the ellipse (…)

15. Then in the **Test Workflow** window, click the ellipse (…) and click **Edit Item**, as shown below:

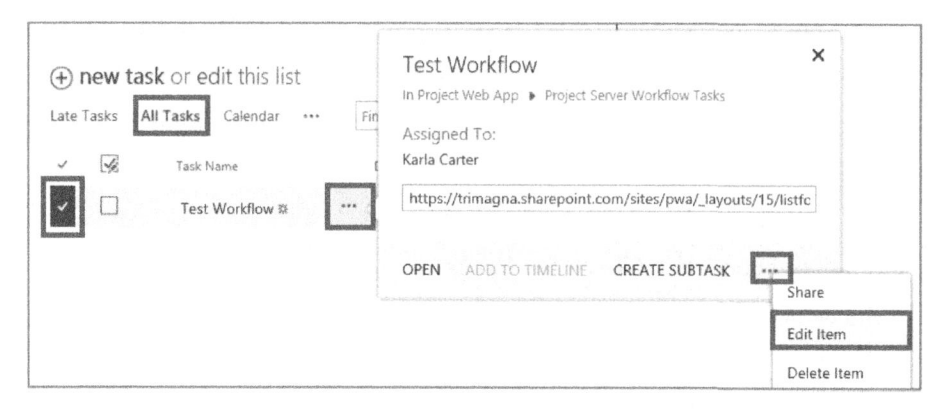

16. At bottom of the **Edit** page, click **Approved**.

17. In the **Project Server Workflow Tasks** page, notice that **Test Workflow** is 100% complete by the checkbox.

18. Sign out as **Karla Carter (KCarter)** from PWA and sign back in to PWA as **Beth Quinlan (BQuinlan)**.

19. From the **PWA Home** page, in the quick launch, click **Projects**.

20. From the **Project Center** page, click **Test Workflow**.

21. On the **Workflow Status** page, expand **All Workflow Stages** at the bottom of the page.

Notice that the **State** column in the **2 – Request Review** stage now displays **Completed** and now **3 – Plan** displays **In Progress (Waiting for input)**.

Now Beth Quinlan must create a project schedule.

22. In the quick launch, under **Test Workflow**, click **Schedule**.
23. Under the **Task Name** column, type **Task 1** and press **Enter**.
24. Type **Task 2** and press **Enter**.
25. Type **Task 3** and press **Enter**.
26. Select all three tasks and from the ribbon, click the **Link Tasks** button, as shown below:

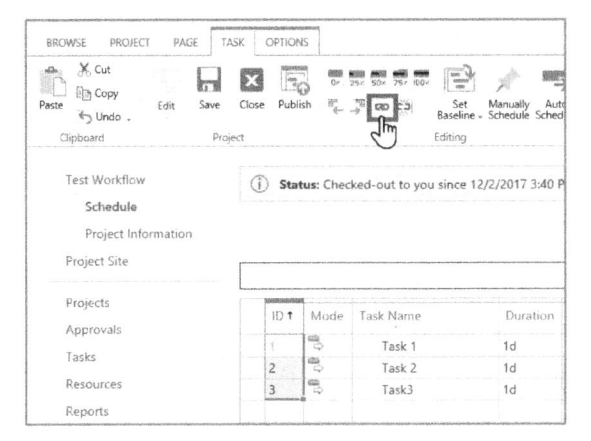

27. From the ribbon, click **Publish.**

> **NOTE:** *At this point we will assume that the planning of the project schedule is finished.*

28. From the ribbon, in the **Editing** section, click **Set Baseline**, as shown below:

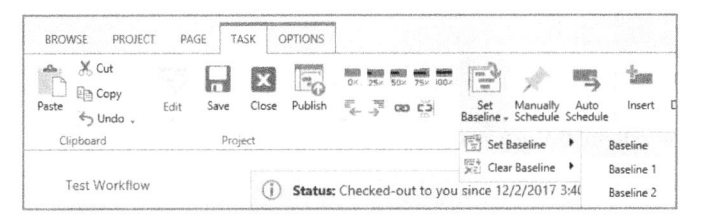

29. From the ribbon, click **Publish.**
30. From the ribbon, click the **Project** tab, and in the **Workflow** section, click **Submit**.
31. On the **Workflow Status** page, expand **All Workflow Stages** at the bottom of the page.

    Notice that the **State** column in the **3 – Plan** stage now displays **Completed** and now **4 – Execute** displays **In Progress (Waiting for input)**.

32. Sign out as **Beth Quinlan (BQuinlan)** from PWA and close the (private mode) web browser.

Lesson 2: Sharing Project Online with External Users

- Overview of SharePoint B2B Sharing
- Configuring Sharing on PWA Site Collection
- Configuring Sharing on Project Web App
- Adding External Users in PWA
- Assigning Project Online Licensing for External Users

In this lesson, you will learn how to share Project Online as a business-to-business (B2B) extranet solution. You will learn how to configure sharing on the PWA site collection and the PWA site. You will also learn how to add external users in PWA and how to assign Project Online licensing for External Users.

Objectives

After completing this lesson, you will be able to:

- Understand SharePoint B2B Sharing
- Configure Sharing on PWA Site Collection
- Configure Sharing on Project Web App
- Add External Users in PWA
- Assign Project Online Licensing for External Users

Overview of Project / SharePoint Online B2B Sharing

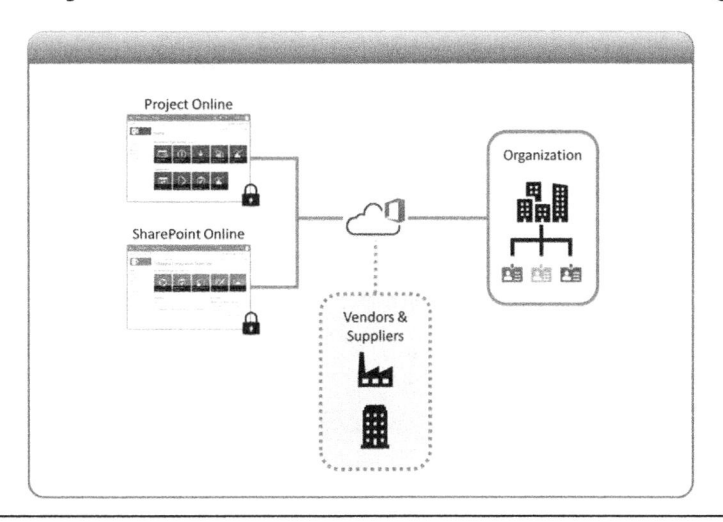

An extranet site in Project Online is a site that you create to allow external partners access to specific content, and to collaborate with them. Extranet sites are a way for partners to securely do business with your organization. Traditionally, deploying a Project Server on-premises extranet site involves complex configuration to establish security measures and governance, including granting access inside the corporate firewall, and expensive initial and on-going cost.

But with Office 365, partners connect directly to a PWA site in Project Online, without access to your on-premises environment or any other SharePoint Online sites. Office 365 Extranet sites can be accessed anywhere there's an Internet connection.

Project / SharePoint Online collaboration features

Allow users to Invite new partner users – In certain site collections, admins can optionally allow users to invite new partner users. In this model, an email invite is sent to the partner user and the user must redeem that invite to access the resource. See Manage external sharing for your SharePoint Online environment for details.

Sharing by site owners only – Ability to have site collections where only site owners can bring in or share with new users. Site members, who are typically external partner users, can see only the existing site members in the site. This helps in governing what partners can see and with whom they can share documents.

Restricted domains sharing – Admins can control the list of partner domains that their employees can share with outside the organization. Either an allow list of email domains or a deny list of email domains can be configured. See Restricted Domains Sharing in O365 SharePoint Online and OneDrive for Business more details.

Auditing & Reporting – Activities of the business partner users are audited, and reports can be viewed in Office 365 Activity Reports.

Configuring Sharing on PWA Site Collection

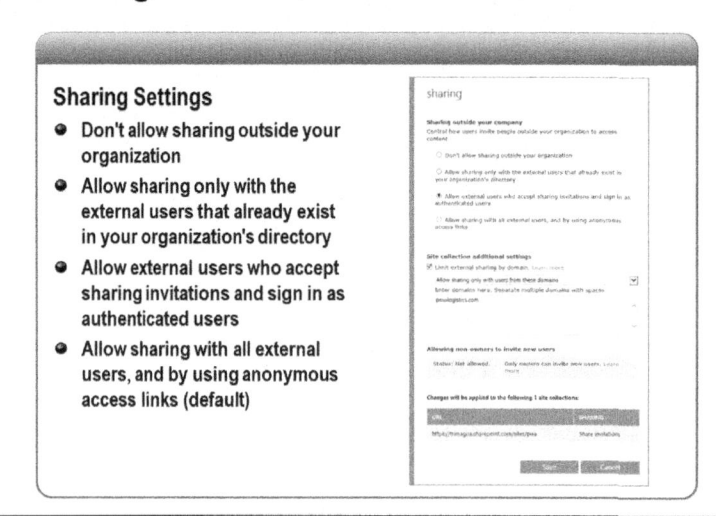

Sharing Settings
- Don't allow sharing outside your organization
- Allow sharing only with the external users that already exist in your organization's directory
- Allow external users who accept sharing invitations and sign in as authenticated users
- Allow sharing with all external users, and by using anonymous access links (default)

Remember that Project Online is a SharePoint application and as such, it follows the same security rules as SharePoint Online. With that being said, SharePoint Online has both global (tenant-wide) and site collection settings for external sharing. The tenant-level settings override any settings at the site collection level. For your Project / SharePoint Online tenant and for each individual site collection, you can choose from the following basic sharing options:

No external sharing – Sites and documents can only be shared with internal users in your Office 365 subscription.

Sharing only with external users in your directory – Sites, folders, and documents can only be shared with external users who are already in your Office 365 user directory. For example, users who have previously accepted a sharing invitation or users who you have imported from another Office 365 or Azure Active Directory tenant.

Sharing with authenticated external users – Sites, folders, and documents can be shared with external users who have a Microsoft account or a work or school account from another Office 365 subscription or an Azure Active Directory subscription.

Sharing with anonymous users – Documents and folders (but not sites) can be shared via an anonymous link where anyone with the link can view or edit the document, or upload to the folder.

How to Configure Sharing for Authenticated External Users

1. Open a web browser and go to the Office 365 Admin center.
2. Sign in as the Office 365 Administrator.
3. On the **Office 365 Admin center – Home** page, in the navigation menu, click **Admin centers → SharePoint**.
4. On the **SharePoint admin center** page, from **Site Collections**, select the check box for the PWA site
5. In the menu bar, click **Sharing**.

6. In the **sharing** window, under **Sharing outside your company**, select **Allow external users who accept sharing invitation and sign in as authenticated users**.

7. Under **Site collection additional settings**, select the check box for **Limit external sharing by domain** and select **Allow sharing only with users from these domains**.

8. In the **Enter domain here** box, type the domain name of your external user account.

9. Click **Save**.

Configuring Sharing on Project Web App

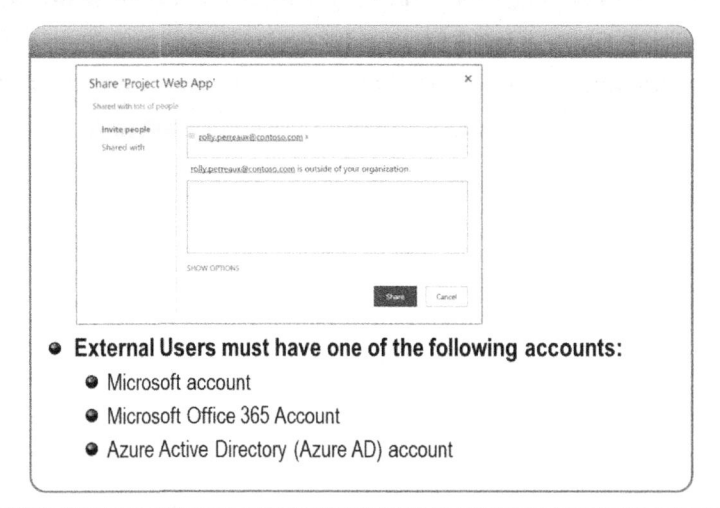

- **External Users must have one of the following accounts:**
  - Microsoft account
  - Microsoft Office 365 Account
  - Azure Active Directory (Azure AD) account

When you share the PWA site with external users, you send them an invitation that they can use to log in to your site. You can send this invitation to any email address. When the recipient accepts the invitation, they log in using either a Microsoft account or a work or school account (Office 365 or Azure AD).

NOTE: You need to be a Site Owner or have full control permissions to share a site with external users.

How to Share Project Web App

1. Sign in as the PWA Administrator.
2. In the upper right corner of the PWA Home page, click **Share**.
3. In the Share dialog box, type the names of the external users you want to invite. For example: someone@outlook.com or someone@contoso.com.

   NOTE: You can also use the Share command to grant internal licensed users access to a site. If you want to do this, just type the names of the people you want to invite.

4. Type a personal message to include with the invitation.
5. Click **Show Options**.
6. By default, an email invitation is sent automatically. Clear the checkbox, **Send an email invitation**, to turn off the email invitation.
7. In the **Select a permission level** list, accept the default **Team Members for Project Web App [Contribute]**.

   NOTE: Do not change the permission level here as we will add the external user to a PWA Group as they will be assigned the group's permission level.

8. Click **Share**.

By default, invitations sent to external users will expire in 90 days. If an invitee does not accept the invitation within 90 days, and you still want that person to have access to your site, you'll need to send a new invitation.

When the external users receive their invitations, they select a button that takes them to a page where they sign into your SharePoint Online site by using a Microsoft account or a work or school account. If users don't have at least a Microsoft account, they can sign up for a free Microsoft account.

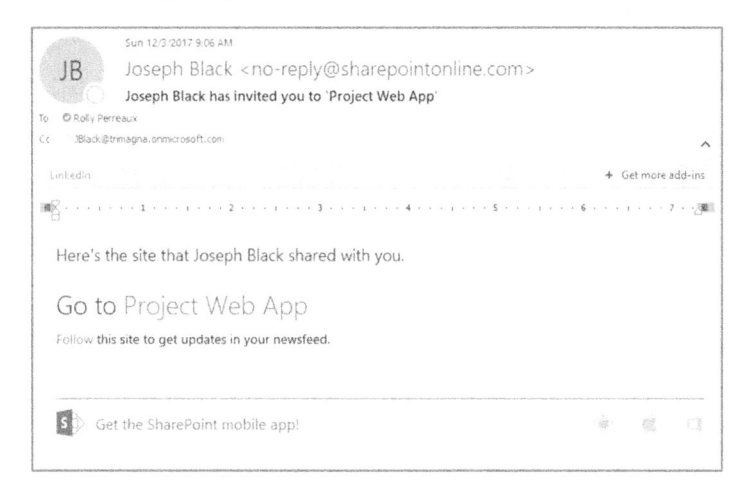

Adding External Users in PWA

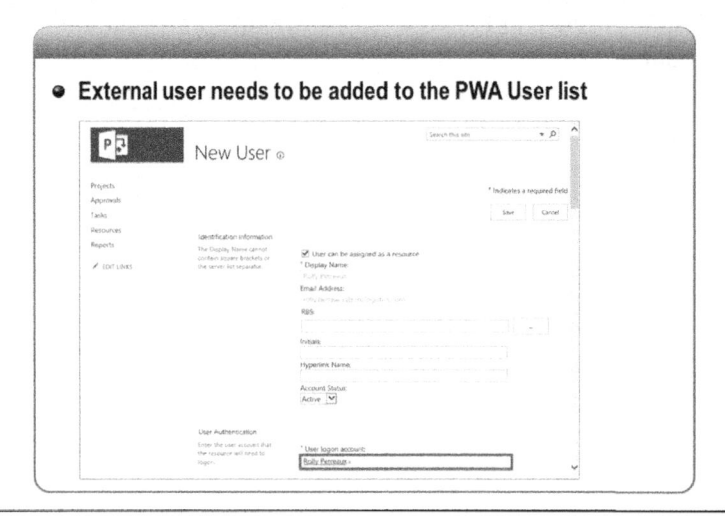

> • External user needs to be added to the PWA User list

External users must also be added as PWA Users. If they are not added, they will receive the following error message when they click the link provided in the invitation email:

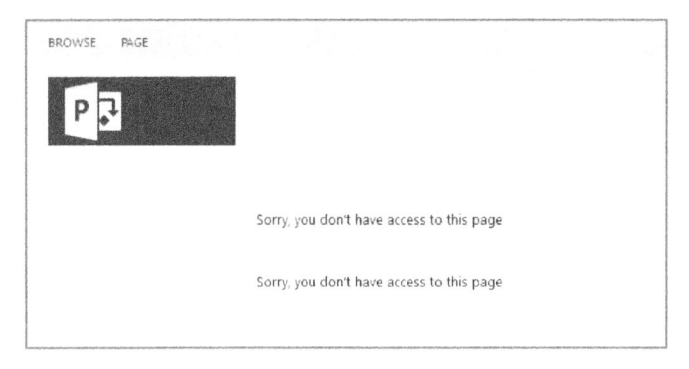

How to Add an External User to PWA

1. Open a web browser and go to the Project Web App site.
2. Sign in as the PWA Administrator.
3. On the **PWA Home** page, click **Settings → PWA Settings**.
4. On the **PWA Settings** page, under the **Security** section, click **Manage Users**.
5. On the **Manage Users** page, in the menu bar, click **New User**.
6. On the **New User** page, under **User Authentication**, in the **User logon account**, type the external user's account and click **Save**.
7. On the **Manage Users** page, review the user list to verify the external user is now listed.

Assigning Project Online Licensing for External Users

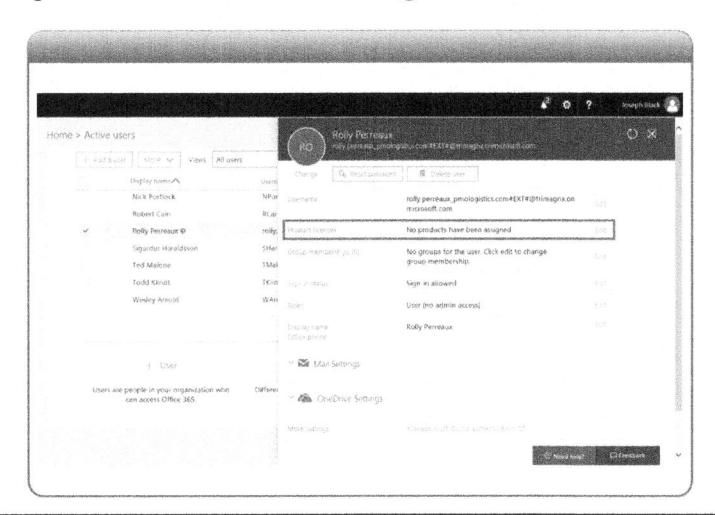

External users must also be assigned a Project Online license. Without this license, the user will receive the following error message:

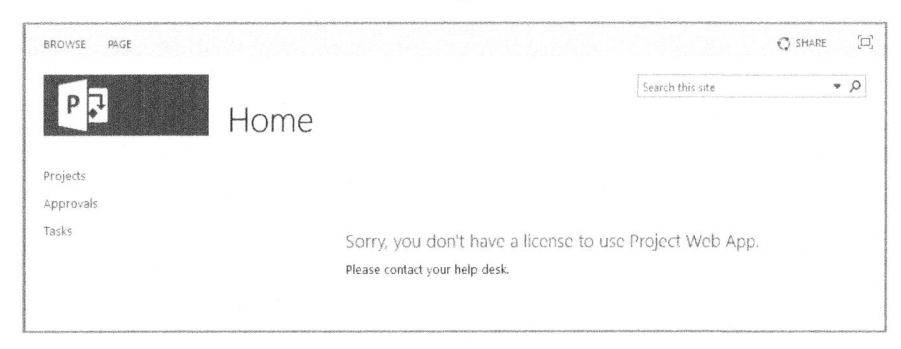

How to Assign Project Online License to an External User

1. Open a web browser and go to the Office 365 Admin center.
2. Sign in as the PWA Administrator.
3. On the **Office 365 Admin center – Home** page, in the navigation menu, click **Users → Active users**.
4. On the **Home > Active users**, select the check box for name of the external user.
5. In the user pane, under **Product licenses**, click **Edit**.
6. In the **Product licenses** pane, select the Project Online license and slide the switch to **On**.
7. Click **Save**, and then click **Close** twice.

NOTE: The external user must sign out and sign back in to the PWA site in order for the Project Online license to take effect.

Practice: Sharing Project Online with External Users

In this practice, you will:
- Configure Sharing on PWA Site Collection
- Configure Sharing on Project Web App
- Add External Users in PWA Users List
- Assign a Project Online Licensing for External User
- Test Access for External User

In this practice, you will share Project Online with external users. You achieve this by configuring sharing on the PWA site collection, then sharing access to the PWA site by sending an invitation to an external user. You will then add the external user as a PWA user and then assign a Project Online license to the external user using the Office 365 Admin center.

NOTE: You must have an Office 365 or Azure AD user account from another domain in order to successfully complete this practice.

Exercise 1: Configuring Sharing on PWA Site Collection

In this exercise, you will configure sharing on the PWA site collection by using the SharePoint admin center.

1. Open a web browser and go to the Office 365 Admin center.
2. Sign in as the PWA Administrator.
3. On the **Office 365 Admin center – Home** page, in the navigation menu, click **Admin centers → SharePoint**.
4. On the **SharePoint admin center** page, from **Site Collections**, select the check box for the PWA site
5. In the menu bar, click **Sharing**.
6. In the **sharing** window, under **Sharing outside your company**, select **Allow external users who accept sharing invitation and sign in as authenticated users**.
7. Under **Site collection additional settings**, select the check box for **Limit external sharing by domain** and select **Allow sharing only with users from these domains**.
8. In the **Enter domain here** box, type the domain name of your external user account.
9. Click **Save**.

Module 7: Project Online Administration

Exercise 2: Configuring Sharing on Project Web App

In this exercise, you will share access to PWA site with your external user account.

10. Open a web browser and go to the PWA site.
11. Sign in as the PWA Administrator.
12. In the upper right corner of the **PWA Home** page, click **Share**.
13. In the **Share** dialog box, type your external user account.
14. Type a personal message to include with the invitation.
15. Click **Share**.

Exercise 3: Adding External Users in PWA Users List

In this exercise, you will add your external user account to the PWA Users listing.

1. On the **PWA Home** page, click **Settings → PWA Settings**.
2. On the **PWA Settings** page, under the **Security** section, click **Manage Users**.
3. On the **Manage Users** page, in the menu bar, click **New User**.
4. On the **New User** page, under **User Authentication**, in the **User logon account**, type your external user account and click **Save**.
5. On the **Manage Users** page, review the user list to verify the external user is now listed.

Exercise 4: Assigning a Project Online License for External User

In this exercise, you will assign a Project Online License for your external user account.

1. Switch to the **Office 365 Admin center – Home** page, in the navigation menu, click **Users → Active users**.
2. On the **Home > Active users**, select the check box for your external user.
3. In the user pane, under **Product licenses**, click **Edit**.
4. In the **Product licenses** pane, select the Project Online license and slide the switch to **On**.
5. Click **Save**, and then click **Close** twice.

Exercise 5: Testing Access for External User

In this exercise, you will test if your external user account has access to the PWA site.

1. Start a Private Mode web browser session and go to the PWA site and sign in as your external user
2. If you have access to the **PWA Home** page, then you have successfully configured external sharing.

Troubleshooting the Practice

If you receive the following error, then you should review the sharing Site Collection settings.

> **That didn't work**
> We're sorry, but rolly.perreaux@contoso.com can't be found in the trimagna.sharepoint.com directory. Please try again later, while we try to automatically fix this for you.

If you receive the following error, then you should review the PWA users list to verify your external user account is listed.

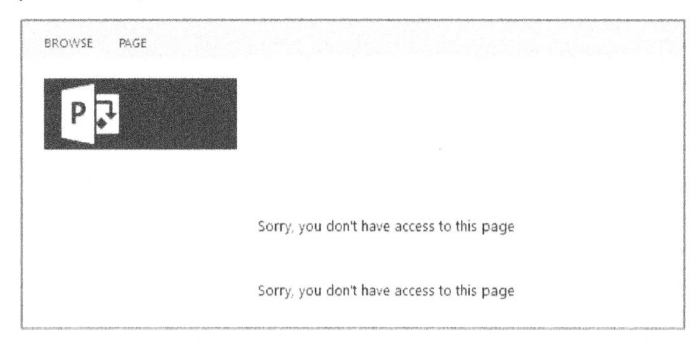

Sorry, you don't have access to this page

If you receive the following error, then you should review the Office 365 users list to verify your external user account has been assigned a Project Online license.

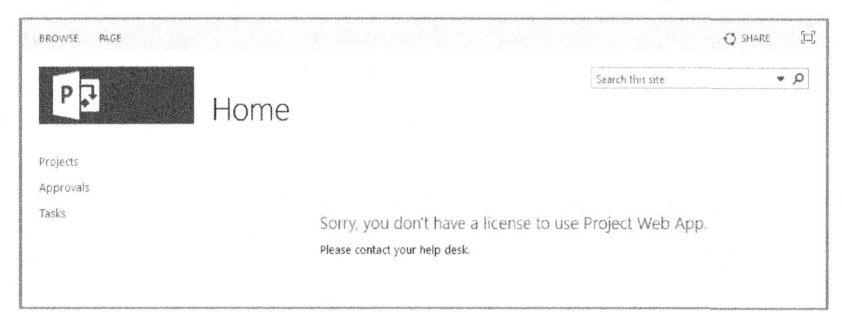

Sorry, you don't have a license to use Project Web App. Please contact your help desk.

Lesson 3: Managing Queue Jobs and Enterprise Objects

- Manage Queue Jobs
- Deleting Enterprise Objects
- Forcing Check-In Enterprise Objects

In this lesson, you will learn how the Project Server Queue works and how to manage queue jobs. You will then learn about the Project Server enterprise objects you can manage.

Objectives

After completing this lesson, you will be able to:

- Manage Queue Jobs
- Delete Enterprise Objects
- Force Check-in of Enterprise Objects

Managing Queue Jobs

The Manage Queue Jobs page allows you to view all the Project Web App Jobs that have been processed by the queue system. You can use the configuration options to filter jobs and only see the jobs that you are interested in viewing. You can also retry or cancel jobs through this page.

How to Access the Manage Queue Job Settings

1. From the PWA site, click **Settings** → **PWA Settings**.
2. On the **PWA Settings** page, in the **Queue and Database Administration** section, click **Manage Queue Jobs**.

NOTE: By default, all job completion states are listed, except Success. To display all successful queue jobs, you must need to add Success to the Selected Job States list.

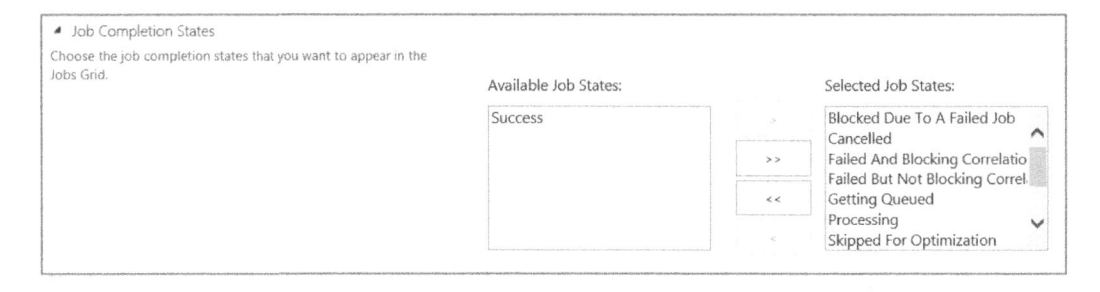

Deleting Enterprise Objects

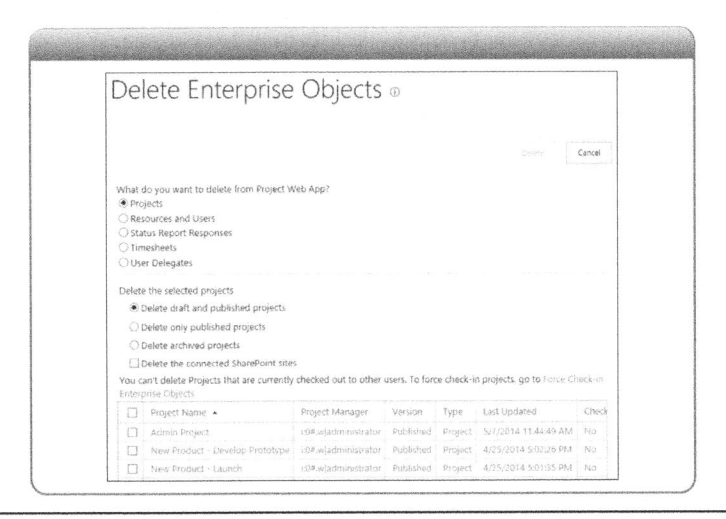

Periodically you may want to create more space in Project Web App by deleting old data about projects, resources, tasks and assignments, task updates, status reports, and timesheets.

How to Delete Enterprise Objects

1. From the PWA site, click **Settings** → **PWA Settings**.

2. On the **PWA Settings** page, in the **Queue and Database Administration** section, click **Delete Enterprise Objects**.

3. In the **What do you want to delete from Project Web App?** section, select the type of data that you want to delete.

    The data that is associated with the data type that you selected appears in the table at the bottom of the page. The table does not appear if you selected Status Report Responses or Timesheets.

4. Select the item that you want to delete in the table and click **Delete**.

    If you selected Status Report Responses or Timesheets, specify the date range for these two items by using the date pickers, and then click **Delete**.

Forcing Check-In of Enterprise Objects

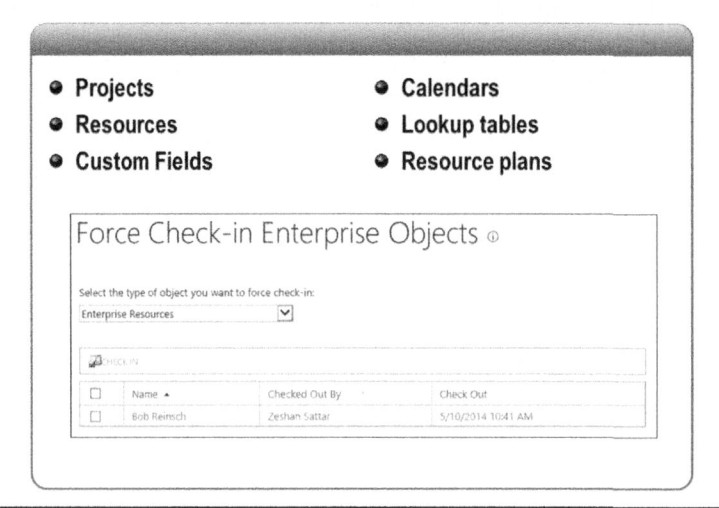

On occasion there may be times when a team member may leave an enterprise object in a state such that he/she cannot check it back into Project Web App. As an administrator, you can force a check-in of the following enterprise objects:

- Projects
- Resources
- Fields
- Calendars
- Lookup tables
- Resource plans

How to Force Check-in of Enterprise Objects

1. From the PWA site, click **Settings → PWA Settings**.
2. On the **PWA Settings** page, in the **Database Administration** section, click **Force Check-in Enterprise Objects**.
3. In the **Select the type of object you want to force check-in** list, click the type of objects that you want to display in the table.
4. Select the object and click **Check In**.

NOTE: When you check in an object that is checked out, the user who has the object checked out can save changes to the database only if he or she saves the object as a new object. Otherwise, the user needs to check out the object again and then make the same changes.

Practice: Managing Enterprise Objects

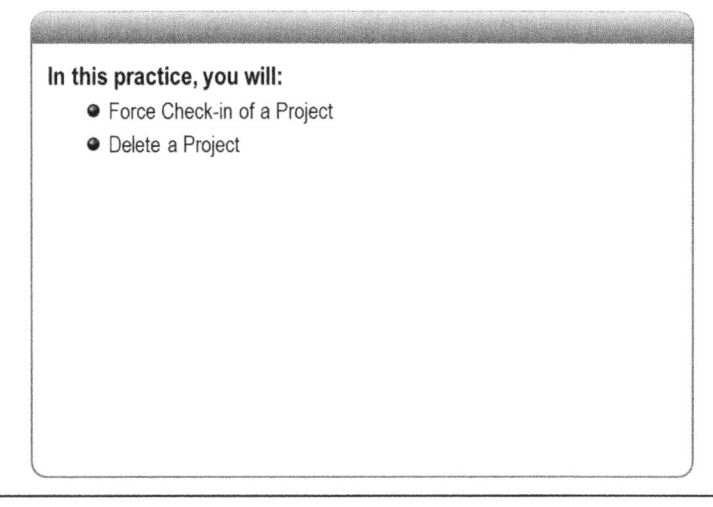

In this practice, you will:
- Force Check-in of a Project
- Delete a Project

In this practice, you will work with enterprise objects.

Exercise 1: Force Check-in of a Project

In this exercise you will force check-in of a checked out project.

1. Switch to the **Project Web App** tab and in the Quick Launch, click **Projects**.
2. On the **Project Center** page, click on **Admin Project**.
3. On the **Schedule: Admin Project** page, from the **Task** tab, click **Edit → In Browser**.

   This checks-out the project.

4. From the PWA site, click **Settings → PWA Settings**.
5. On the **PWA Settings** page, in the **Queue and Database Administration** section, click **Force Check-in Enterprise Objects**.
6. On the **Force Check-in Enterprise Objects** page, select the check box for **Admin Project** and click **Check In**.
7. In the **Message from webpage** dialog box, click **OK**.
8. On the **Force Check-in Enterprise Objects** page, press the **F5** key to refresh the page.

   Notice the Job State column displays Processing. Press F5 again until the project is longer checked-out.

Exercise 2: Delete a Project

In this exercise, you will delete **Admin Project** from the project listing.

1. From the PWA site, click **Settings → PWA Settings**.
2. On the **PWA Settings** page, in the **Queue and Database Administration** section, click **Delete Enterprise Objects**.
3. On the **Delete Enterprise Objects** page, ensure that **Projects** and **Delete draft and published projects** are selected.

4. Select the **Delete the connected SharePoint sites** check box, then select the **Admin Project** check box and click **Delete**.

5. In the **Message from webpage** dialog box, click **OK**.

   Notice the Job State column displays Processing. Press F5 again until the project is deleted.

6. On the **Delete Enterprise Objects** page, click **Cancel**.

Lesson 4: Troubleshooting Resources

- Office 365 Service Health
- Office 365 Message Center
- Office 365 Usage Reports
- Tune Project Online Performance

In this lesson, you will learn the various troubleshooting resources available to assist a PWA Administrator.

Objectives

After completing this lesson, you will identify:

- Office 365 Service Health
- Office 365 Message Center
- Office 365 Usage Reports
- Tune Project Online Performance

Office 365 Service Health

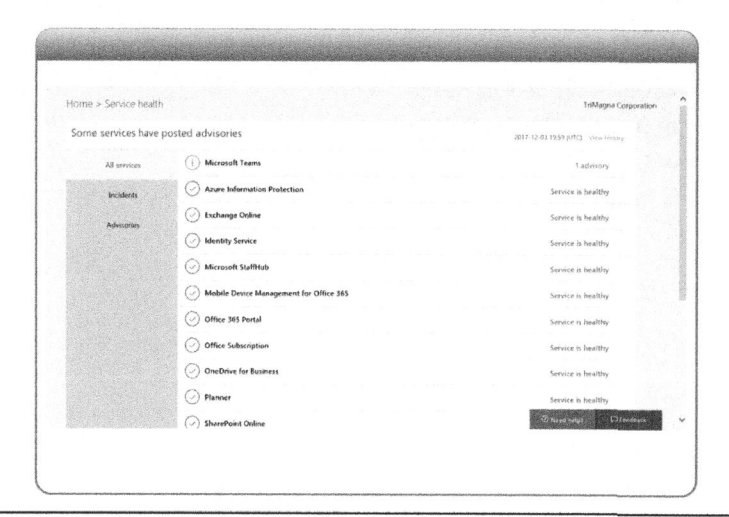

You can view the health of all Office 365 cloud services on the Office 365 Service health page in the admin center. If you are experiencing problems with a cloud service, you can check the service health to determine whether this is a known issue with a resolution in progress before you call support or spend time troubleshooting.

You can also quickly see at a glance the Service health notifications on the Office 365 Home page in the Service health card.

Most of the time services will appear as healthy with no further information. However, when a service is experiencing a problem, the issue is identified as either an advisory or an incident and shows a current status.

NOTE: Planned maintenance events aren't shown in service health. You can track planned maintenance events by staying up to date with the Message center.

Advisory

If a service has an advisory shown, we are aware of a problem that is affecting some users, but the service is still available. In an advisory, there is often a workaround to the problem and the problem may be intermittent or is limited in scope and user impact.

Incident

If a service has an active incident shown, it's a critical issue and the service or a major function of the service is unavailable. For example, users may be unable to send and receive email or unable to sign-in. Incidents will have noticeable impact to users. When there is an incident in progress, we will provide updates regarding the investigation, mitigation efforts, and confirmation of resolution in the Service health dashboard.

Office 365 Message Center

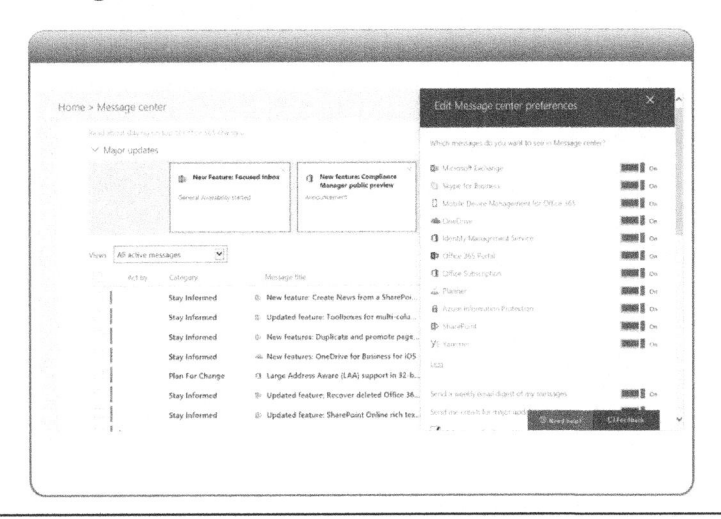

Visit the Message center to keep track of upcoming feature releases or issues. Official announcements about new and changed features are posted here to help you to take a proactive approach to change management. Each post gives you a high-level overview of a planned change and how it may affect your users, and links out to more detailed information to help you prepare.

To open Message center, in the Office 365 admin center, go to **Health** > **Message center**, or click the Message center card on the **Home** dashboard. The dashboard card always displays the last three messages that we posted and links to the full Message center page.

To Set Your Message Center Preferences

If administration is distributed across your organization, you may not want or need to see posts about all Office 365 services. Each administrator can set preferences that control which messages are displayed in Message center. To ensure you never miss an important Office 365 services post you can opt-in to receive a weekly email digest of messages.

1. Select **Edit Message center preferences** at the top of Message center.

2. Make sure that the toggle is set to **On** for each service that you want to monitor. Click the toggle to change the setting to **Off** for the services you want to filter out of your Message center view.

3. If your admin account is enabled for the **First Release** program, you can choose to have a weekly summary of Message center posts delivered to your email. Select **Send a weekly email digest of my messages**, then enter the email addresses where you want to receive the digest. You can enter the email address for an Office 365 group or a distribution list if more than two people should get the digest email. Digest emails are turned on by default.

4. Click **Save** to keep your changes.

Office 365 Usage Reports

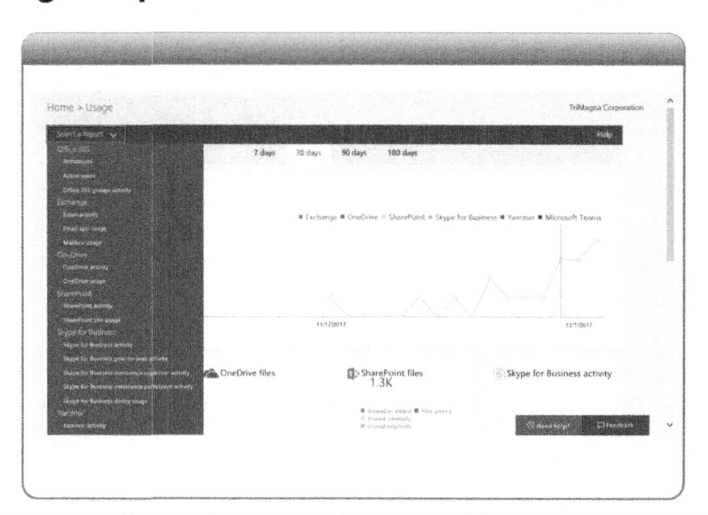

You can easily see how people in your business are using Office 365 services. For example, you can identify who is using a service a lot and reaching quotas, or who may not need an Office 365 license at all.

Reports are available for the last 7, 30, 90, and 180 days. Data won't exist for all reporting periods right away. The reports become available with a minimum 30 days of data.

How to View the Reports Dashboard

1. From the **Office 365 admin center** page, choose **Reports > Usage**.
2. Choose **Select a report** at the top of the dashboard to select from a list of all available reports. Or, click an at-a-glance activity widget for a service (email, OneDrive, etc) to view more information.

Who Can View Office 365 Reports

People who have the following permissions:

- Office 365 global admins
- Exchange admins
- SharePoint admins
- Skype for Business admins
- Reports reader

Tune Project Online Performance

- **Best Practices of the following topics at:** http://aka.ms/tunepo
  - Permissions Modes: SharePoint or Project
  - Enterprise Project Types
    - Project site configuration
    - Synchronization mechanisms between Project Online and SharePoint Online
  - Active Directory Resource Pool Sync
  - PWA Pages and Views Customizations
  - Custom Project Detail Pages and Workflows
  - OData and Reporting
  - Project Online Quota

With the launch of Project Online a few years ago, organizations of all sizes have been able to use Microsoft's rich set of Project Portfolio Management (PPM) capabilities within the convenience of our Office 365 cloud infrastructure.

Although one of the obvious benefits of using a cloud-based service is avoiding having to deal with deployment, setup, and hardware and software tuning, there are still some steps you can take to ensure your organization receives the optimum performance out of Project Online.

Project Online offers many configuration and customization settings, but customizations can result in performance impacts. To help you make informed decisions when customizing and configuring Project Online, the following article highlights performance impacts and tradeoffs of some of the most common Project Online settings.

http://aka.ms/tunepo

Best Practices of:

- Permissions Modes: SharePoint or Project
- Enterprise Project Types
- Project site configuration
- Synchronization mechanisms between Project Online and SharePoint Online
- Active Directory Resource Pool Sync
- PWA Pages and Views Customizations
- Custom Project Detail Pages and Workflows
- OData and Reporting
- Project Online Quota

Summary

> **In this module, you learned how to:**
> - Work with Workflows and PDPs
> - Share Project Online with External Users
> - Manage Queue Jobs and Enterprise Objects
> - Use Troubleshooting Resources

In this module, you learned how to work with Project Online Workflows and its corresponding elements. You learned how to share Project Online with external users of your organization. You also learned how to manage queue jobs and enterprise objects and the troubleshooting resources available for a PWA Administrator.

Objectives

After completing this module, you learned how to:

- Work with Workflows and PDPs
- Share Project Online with External Users
- Manage Project Server Queue Jobs and Enterprise Objects
- Use Troubleshooting Resources

Hands-On Lab:
How to Create a Project Online Power BI Center

Contents

Overview 1
Exercise 1: Creating a Modern UI SharePoint Site Collection 2
Exercise 2: Signing Up for a Power BI Account 4
Exercise 3: Using the Power BI Project Online Content Pack 6
Exercise 4: Upgrading Free Power BI account to Power BI Pro 8
Exercise 5: Adding Power BI Reports to a SharePoint Page 8
Exercise 6: Modifying the Power BI Center Home Page 11
Exercise 7: Sharing the Power BI Center Site 13
Exercise 8: Sharing the Power BI Dashboard and Testing 14
Summary 17

EXCLUSIVELY PUBLISHED BY

PMO Logistics
679 Roberta Avenue
Winnipeg, Manitoba, Canada R2K 0K9

Copyright © 2017 by Roland Perreaux

All rights reserved. No part of the contents of this document may be reproduced or transmitted in any form or by any means without written permission of the publisher.

Information in this document, including URL and other Internet Web site references, is subject to change without notice. Unless otherwise noted, the example companies, organizations, products, domain names, e-mail addresses, logos, people, places, and events depicted herein are fictitious, and no association with any real company, organization, product, domain name, e-mail address, logo, person, place, or event is intended or should be inferred. Complying with all applicable copyright laws is the responsibility of the user. Without limiting the rights under copyright, no part of this document may be reproduced, stored in or introduced into a retrieval system, or transmitted in any form or by any means (electronic, mechanical, photocopying, recording, or otherwise), or for any purpose, without the express written permission of PMO Logistics Inc.

The names of manufacturers, products, or URLs are provided for informational purposes only and PMO Logistics makes no representations and warranties, either expressed, implied, or statutory, regarding these manufacturers or the use of the products with any Microsoft technologies. The inclusion of a manufacturer or product does not imply endorsement of Microsoft of the manufacturer or product. Links are provided to third party sites. Such sites are not under the control of PMO Logistics and PMO Logistics is not responsible for the contents of any linked site or any link contained in a linked site, or any changes or updates to such sites. PMO Logistics is not responsible for webcasting or any other form of transmission received from any linked site. PMO Logistics is providing these links to you only as a convenience, and the inclusion of any link does not imply endorsement of PMO Logistics of the site or the products contained therein.

PMO Logistics may have patents, patent applications, trademarks, copyrights, or other intellectual property rights covering subject matter in this document. Except as expressly provided in any written license agreement from PMO Logistics, the furnishing of this document does not give you any license to these patents, trademarks, copyrights, or other intellectual property.

PMO Logistics, Professional Training Series, Upgrading Skills Series, TriMagna Corporation and TriMagna Corporation logo are either registered trademarks or trademarks of PMO Logistics Inc. in Canada, the United States and/or other countries.

Microsoft, Active Directory, Internet Explorer, Outlook, Project Server, SharePoint, SQL Server, Visual Studio, Windows and Windows Server are either registered trademarks or trademarks of Microsoft Corporation in the United States and/or other countries.

All other trademarks are property of their respective owners.

Author: Rolly Perreaux, PMP, MCSE, MCT

Publisher: PMO Logistics
Developmental Editor: Heather Perreaux
Cover Graphic Design: Andrea Ardiles
Technical Testing: Underground Studioworks

Post-Publication:
Errata List Contributors:

Overview

In this Hands-On Lab, you will learn how to create a custom SharePoint Site Collection using Modern UI to host Power BI reports. You will also create a new SharePoint page to be used to host a Power BI report. You will then add the Power BI web part to the page and embed the link provided from Power BI Pro.

You will sign up for a Power BI (free version) and then upgrade to Power BI Pro and will be using the Project Online Content Pack to download the built-in dashboards and reports that will use the data from the Project Online link to PWA.

You will then modify the Power BI Center home page and share the site and Power BI reports & dashboards with members of a security group and a user.

Objectives

After completing this Hands-On Lab, you will be able to:

- Create a Modern UI SharePoint Site Collection
- Sign Up for a Power BI Account
- Use the Power BI Project Online Content Pack
- Upgrade Free Power BI account to Power BI Pro
- Add Power BI Reports to a SharePoint Page
- Modify the Power BI Center Home Page
- Share the Power BI Center Site
- Share the Power BI Dashboard and Testing

Exercise 1: Creating a Modern UI SharePoint Site Collection

In this exercise, you will create a new SharePoint Site Collection that will be used as the foundation of the Power BI Center site.

1. Open a web browser and go to the **Office 365** home page at: https://www.office.com and sign is as the Office 365 Global administrator.

2. On the **Office 365** home page, click the **App Launcher** and click **SharePoint**.

3. On the **SharePoint** home page, click **Create site**.

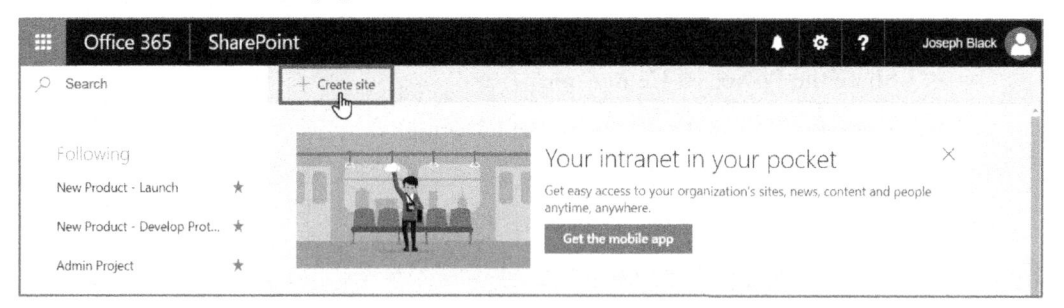

4. In the **Create a site** pane, click **Communication site**.

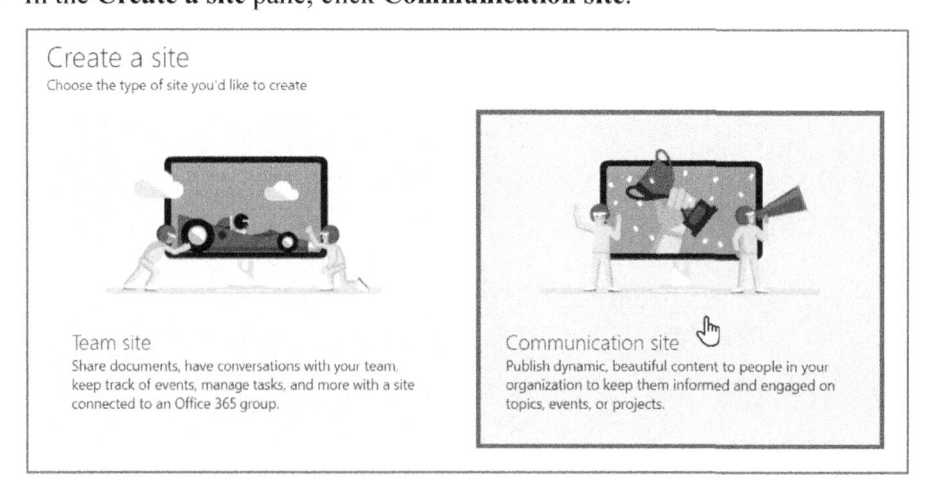

How to Create a Project Online Power BI Center

5. In the **Communication site** pane, complete the form with the following settings:

Setting	Perform the following:
Choose a design	Select **Topic**
Site name	Type **<companyname> Power BI Center**
Site address	Click **Edit** and type **powerbi**.

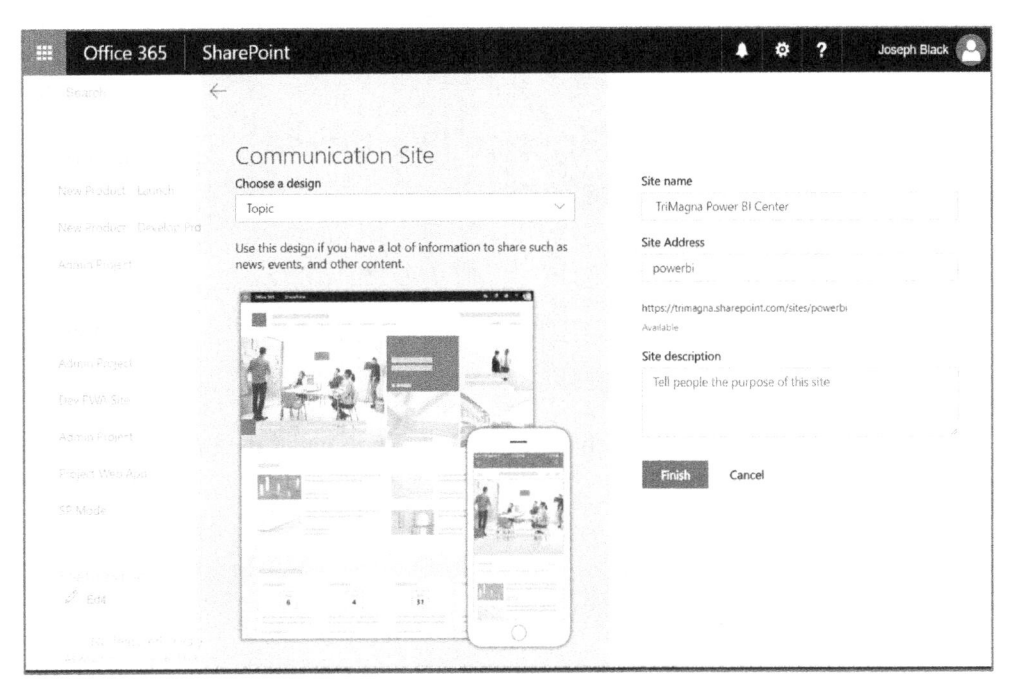

6. Click **Finish**.

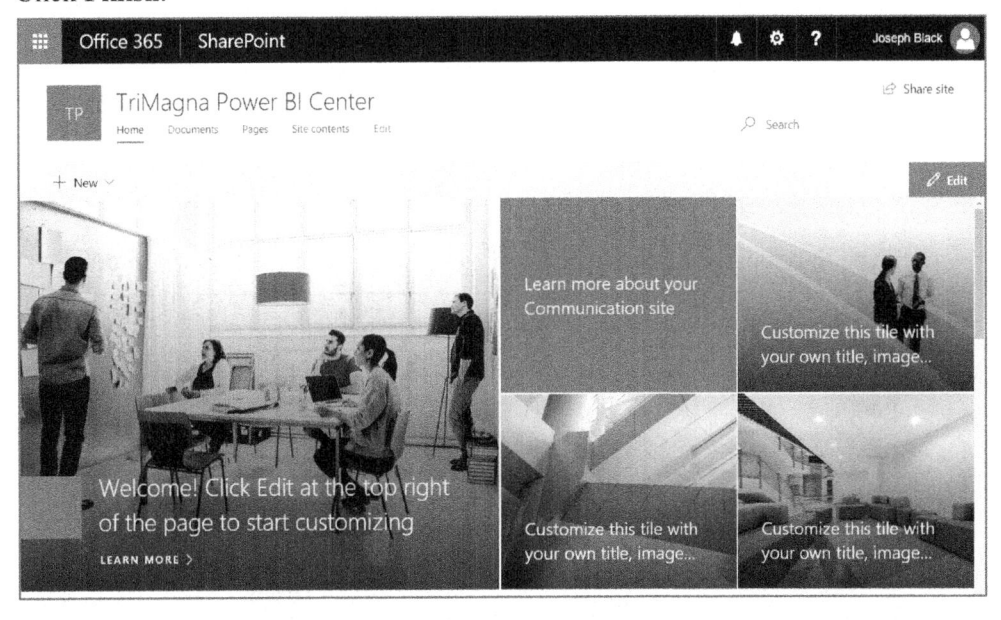

Exercise 2: Signing Up for a Power BI Account

In this exercise, you will sign up for a free Power BI account.

1. Open a new tab on your web browser and go to http://powerbi.com.
2. From the **Microsoft Power BI** web page, click on either **Start Free** OR **Sign up free**.

3. On the **Getting Started** web page, scroll down to under the **Power BI** section and click **Try Free >**.

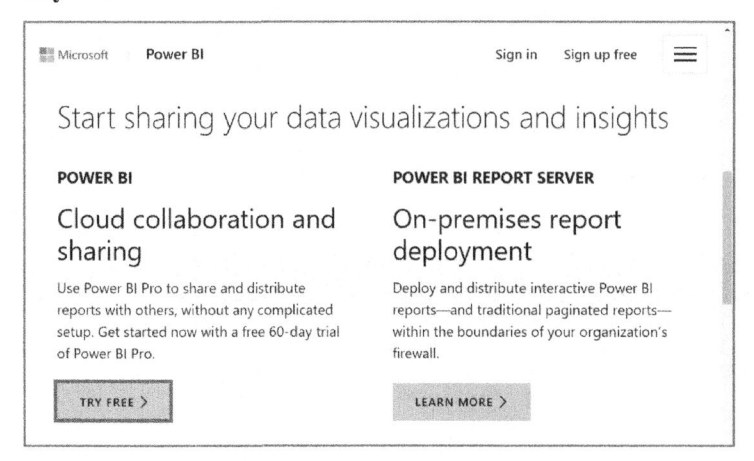

4. Type the email address you are signing up with, and then click **Sign up**.

   NOTE: You must you use a work, or school, email address to sign up. Email addresses such as outlook.com, hotmail.com, gmail.com and others are not allowed

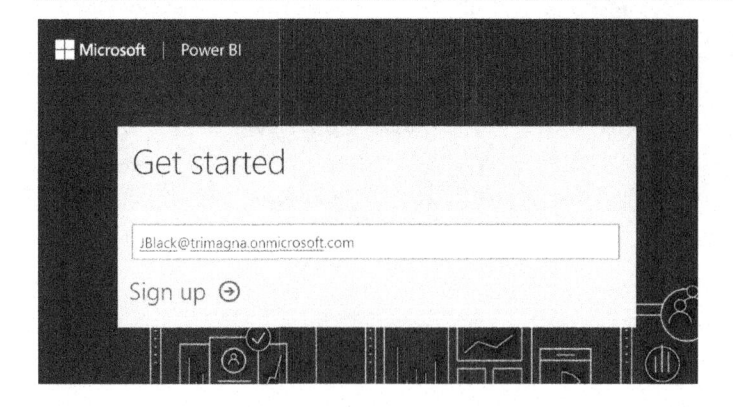

5. You will get a message indicating to check your email. Click on the link in the email validate your email account and complete the instructions from the web page.

> **Great! Go check your email.**
>
> To finish signing up, click the link in the mail from Office 365.
>
> Didn't get the mail? Check your spam folder or resend the mail

However, if you are already using a Microsoft service, such Office 365, you will instead receive a message to Sign in. Sign in with your account and complete the instructions from the web page.

> **You have an account with us**
>
> You're using JBlack@trimagna.onmicrosoft.com with another Microsoft service already. To finish signing up for Microsoft Power BI, sign in with your existing password.
>
> Sign in →

6. After signing in, agree to the terms and conditions by clicking **Start**.

> **Almost there**
>
> You're signed in as JBlack@trimagna.onmicrosoft.com
>
> By choosing **Start**, you agree to our terms and conditions and understand that your name and email address will be visible to other people in your institution. Microsoft Privacy Policy
>
> Start →

7. You will then be taken to https://app.powerbi.com and you can begin using Power BI as a free user.

Exercise 3: Using the Power BI Project Online Content Pack

In this exercise, you will be using the Project Online Content Pack to download the built-in dashboards and reports that will use the data from the Project Online link to PWA.

1. On **Power BI** web page, under **Microsoft AppSource / Services** section, click **Get**.

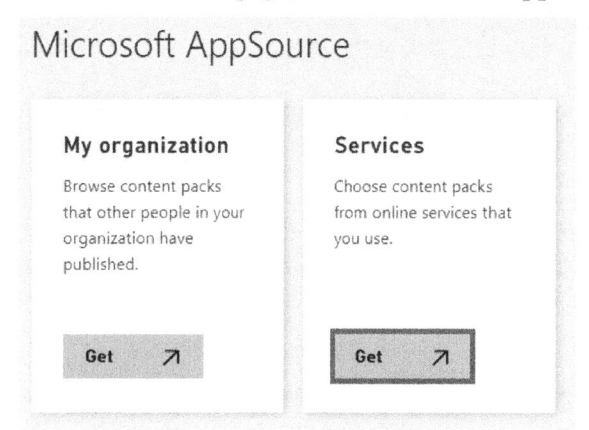

2. In the **AppSource** pane, in the **Search** box, type **Project Online** and click **Get in now**.

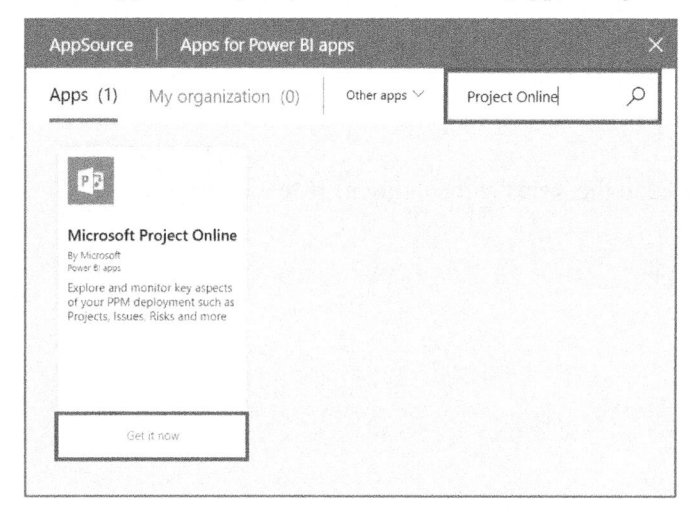

3. In the **Connect to Microsoft Project Online** pane, in the **PWA Site URL** box, type the URL to your production PWA site and click **Next**.

How to Create a Project Online Power BI Center

4. In the **Authentication method** list, select **OAuth2** and click **Sign in**.

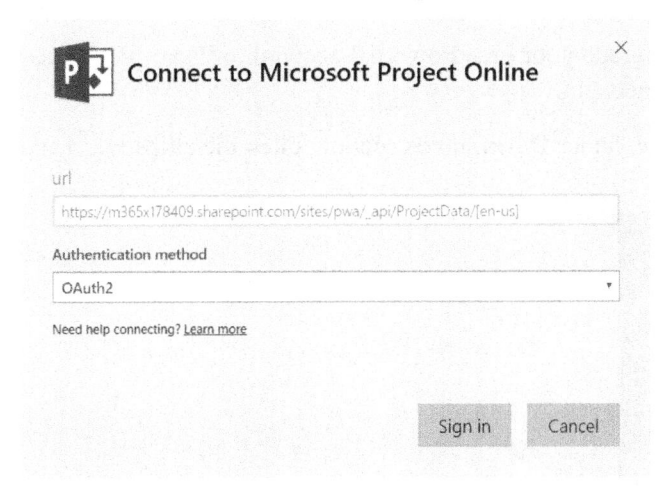

5. Sign in as the PWA Administrator.

    It might take a while for your PWA data to be imported to the Content Pack.

6. When the data has finished being imported, close the **Your dataset is ready!** window.

7. In the navigation pane, expand **My Workspace** and under **Dashboards** or **Reports**, click on **Microsoft Project Online** to review the content.

    NOTE: *Please notice that there are numerous tabs at the bottom of Reports main window.*

Exercise 4: Upgrading Free Power BI account to Power BI Pro

In this exercise you will upgrade your Free Power BI account to Power BI Pro as you cannot share reports with the free account.

1. On **Power BI** web page, under **Dashboards** section, click the ellipse (…) and click **Share**.

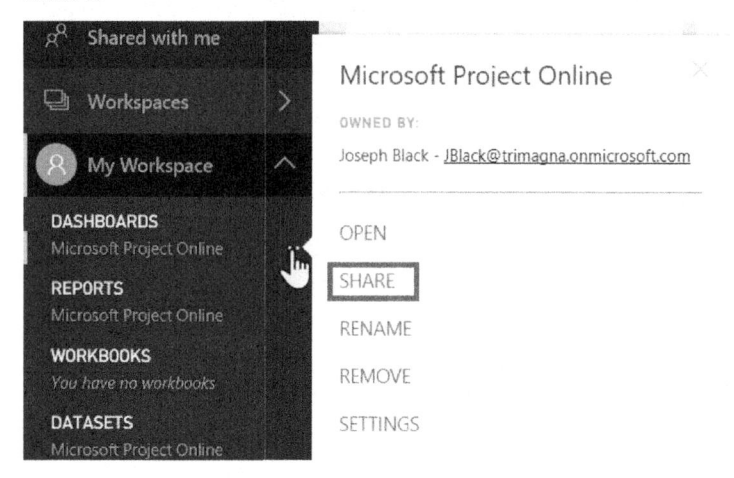

2. In the **Upgrade to Power BI Pro** pane, click **Try Pro for free**.

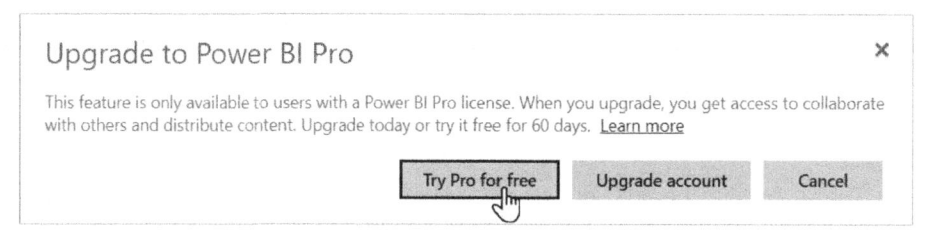

3. In the **Start 60-Day free Pro trial** pane, click **Start trial**.

4. In the **Success! Trial extended** pane, click **Close**.

    The Power BI web page is refreshed.

5. On **Power BI** web page, expand **My Workspace** and under **Reports**, click on **Microsoft Project Online**.

Exercise 5: Adding Power BI Reports to a SharePoint Page

In this exercise you will create a new SharePoint page to be used to host a Power BI report. You will also add the Power BI web part to the page and embed the link provided from Power BI Pro.

1. Switch to the **Power BI Center** page, and from the menu, click **Pages**.

How to Create a Project Online Power BI Center   9

2. On the **Site Pages** page, from the menu bar, click **New** and click **Site Page**.

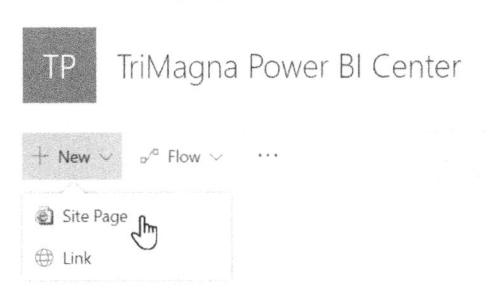

3. On the **Share your ideas** pane, click **Not Now**.
4. Click on **Name your page** and type **Portfolio Report.**
5. In the left top corner, as shown below, click the **X** to remove the banner image.

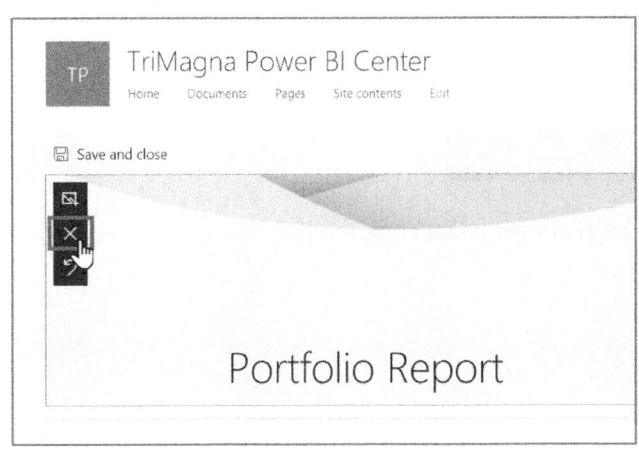

6. Then click the plus sign (**+**) and in the **Search** box, type **Power BI** and then click on **Power BI**.

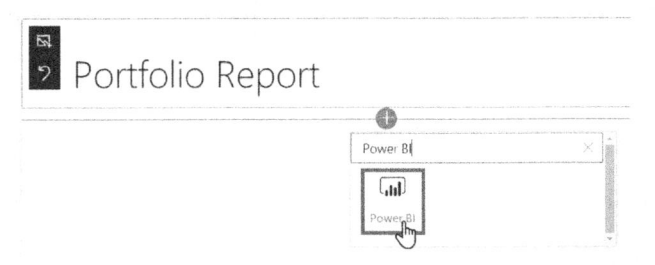

The Power BI web app is added to the page.

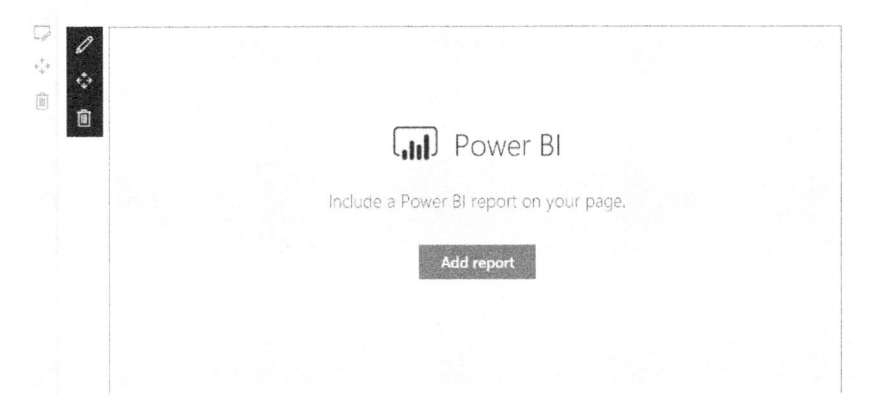

7. Switch to the **Power BI** web page, expand **My Workspace** and under **Reports**, click on **Microsoft Project Online**.
8. Ensure that the **Portfolio Dashboard** report is selected from the bottom tab.
9. Then from the menu, click **File → Embed in SharePoint Online**.

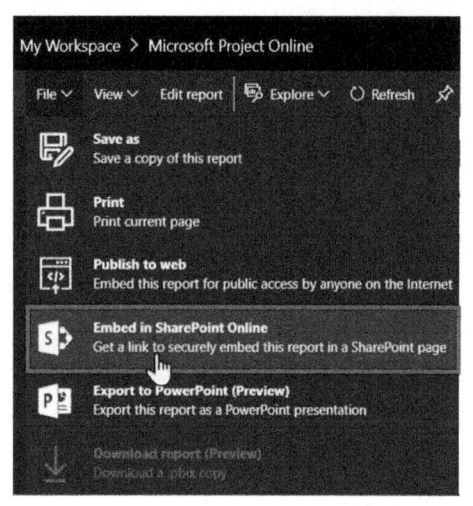

10. In the **Embed link for SharePoint** pane, copy the link that is displayed and then click **Close**.

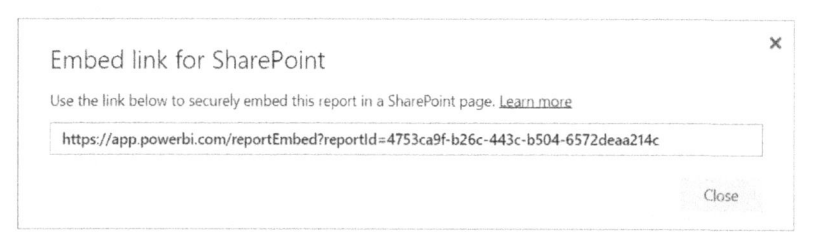

11. Switch tabs to the **Portfolio Report** page, in the **Power BI** web app, click **Add report**.
12. In the **Power BI** side pane, in the **Power BI report link** box, paste the copied link from Power BI, in the **Page name** list, select **Portfolio Dashboard**.

    Notice at the bottom of the Power BI web app, all the Power BI reports are listed.

13. Turn off **Show Navigation Pane**.

How to Create a Project Online Power BI Center 11

Now only the Portfolio Dashboard report is displayed.

14. At the top of the **Portfolio Report** page, click **Save and close**.
15. Notice at the top right of the page, it displays **Draft saved <date>** and a **Publish** button.

    No one will be able to view this page until it is published and shared.

16. Click on **Publish**.

Exercise 6: Modifying the Power BI Center Home Page

In this exercise you will modify the Power BI Center home page by editing the content and link of one of the panels.

1. At the top of the page, from the menu, click **Home**.
2. On the right side of the **Home** page, click **Edit**.

3. Click on **Learn more about your Communication site** and then click **Edit details** icon.

4. In the **Hero** side pane, under **Link**, click **Change**.
5. In the **Recent items** pane, click on **Portfolio Report** and click **Open**.
6. In the **Hero** side pane, in the **Title** box, type **Power BI Portfolio Reports**.
7. Close the **Hero** side pane.

   Continue to make any edits to the page as necessary

8. To make changes to the web part layout, click on the **Edit web part** icon.

9. In the **Hero** side pane, you can test the different layout options that suit your needs. When finished, close the Hero side pane.

   NOTE: *Any changes are automatically applied, but remember they are not published yet.*

10. At the top left side of the **Home** page, click **Save and close**.
11. At the top right side of the **Home** page, click **Publish**.

Exercise 7: Sharing the Power BI Center Site

In this exercise you will share the Power BI Center site with members of the Executives Grp, Business Analysts Grp security group and with a user.

1. At the top right side of the **Home** page, click **Share site**.

2. On the **Share site** side pane, in the box, type **Executives Grp**, **Business Analysts Grp** and **Beth Quinlan**. (ensure they only have **Read** permission)

3. On the **Share site** side pane, click **Share**.

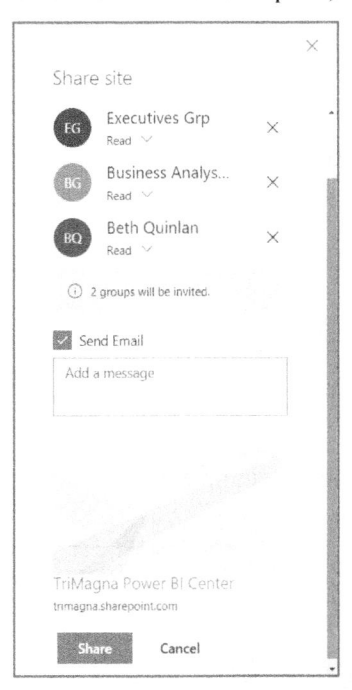

4. On the **Site Pages** page, hover over the **Portfolio Report.aspx** file and click on the check box and then in the menu bar, click **Share**.

Exercise 8: Sharing the Power BI Dashboard and Testing

In this exercise you will start a new private mode and sign in as Ted Malone and attempt to access the Power BI Center site and Portfolio Report page. As Ted, you will need to sign up for a free Power BI account and then upgrade the account to a Power BI Pro account. During this exercise, as the Administrator, you will need to share your Power BI dashboard with security groups and users.

1. Start a new Private Mode session in your web browser and go to: https://<tenant>.sharepoint.com/sites/powerbi.

2. Sign in as **Ted Malone** (tmalone) as he is a member of Executive Grp.

3. On the **Power BI Center Home** page, click on the **Power BI Portfolio Reports** panel.

4. On the **Portfolio Report** page, Ted receives the following message:

   > To view this content, you'll need to sign up for Power BI. Learn more about Power BI.

   This is because he does not have a Power BI Pro account. Only users with Pro accounts can view the Power BI report content.

5. Click on the **Learn more about Power BI** link.

   This will open a new tab.

6. On **Power BI** page, in the top right corner, click **Sign up free**.

7. On the **Getting started...** page, scroll down the page and under **Cloud collaboration and sharing**, click **Try Free >**.

8. In the **Get started** box, type the email address you are signing up with, and then click **Sign up**.

   NOTE: *You must you use a work, or school, email address to sign up. Email addresses such as outlook.com, hotmail.com, gmail.com and others are not allowed*

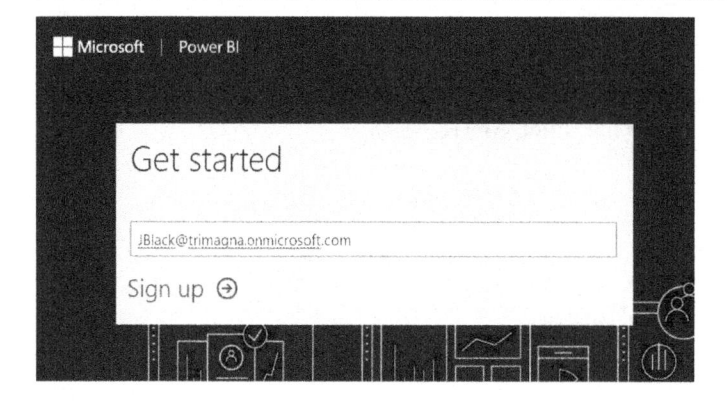

   In the **You have an account with us** box, click **Sign in**.

9. In the **Almost there** box, click **Start**.

10. In the **Invite more people** box, click **Skip**.

11. You will then be taken to https://app.powerbi.com and you can begin using Power BI as a free user.

How to Create a Project Online Power BI Center   15

However, we require Power BI Pro to view the Portfolio Report on the SharePoint site.

12. On the **Power BI** page, in the navigation menu, click **Workspaces** and then click **Create app workspace**.

13. In the **Upgrade to Power BI Pro** window, click **Try Pro for free** and then click **Start trial**.

14. In the **Success! Trial extended** window, click **Close**.

15. Switch over to the **Portfolio Report** tab and refresh the web page.

    The Power BI web part checks Ted Malone's permissions

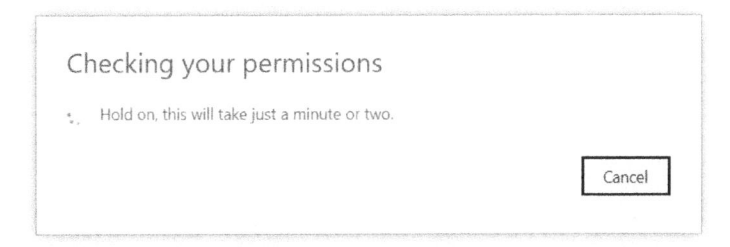

16. On the **Portfolio Report** page, Ted receives the following message:

    This is because he has not been assigned permission to view the Power BI report content from the owner of the report.

17. Switch back to the web browser where you are signed in as the Office 365 Global administrator and select the **Power BI Pro** tab.

18. On the **Power BI** page, under **My Workspace** > **Dashboards**, click the ellipse (**...**) and then click **Share**.

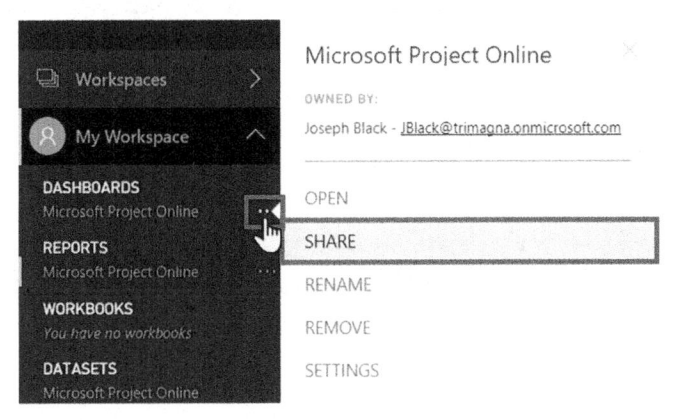

19. In the **Share dashboard** pane, in the **Grant access to** box, type **Executives Grp**, **Business Analysts Grp** and **Beth Quinlan**.

20. Clear the check boxes, **Allow recipients to share your dashboard** and **Send email notifications to recipients**, then click **Share**, as shown below:

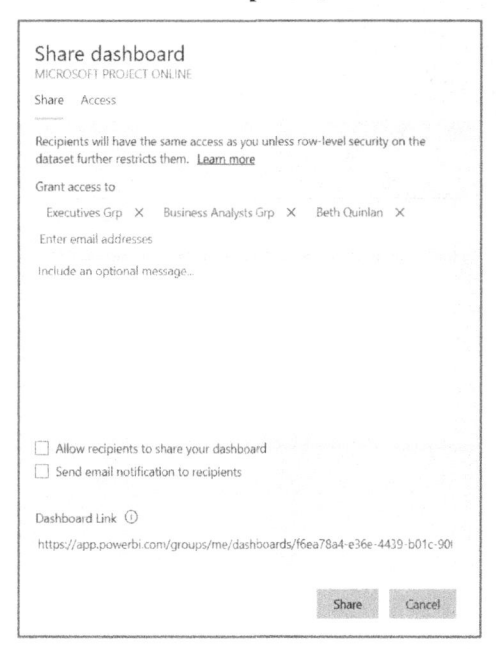

NOTE: In this example we are sharing the dashboard because it also includes data from the Portfolio Dashboard report.

21. Switch back to the web browser where you are signed in as **Ted Malone** on the Power BI Center site.

22. Refresh the **Portfolio Report** page.

   After checking Ted's permission, the Power BI web part displays the report content.

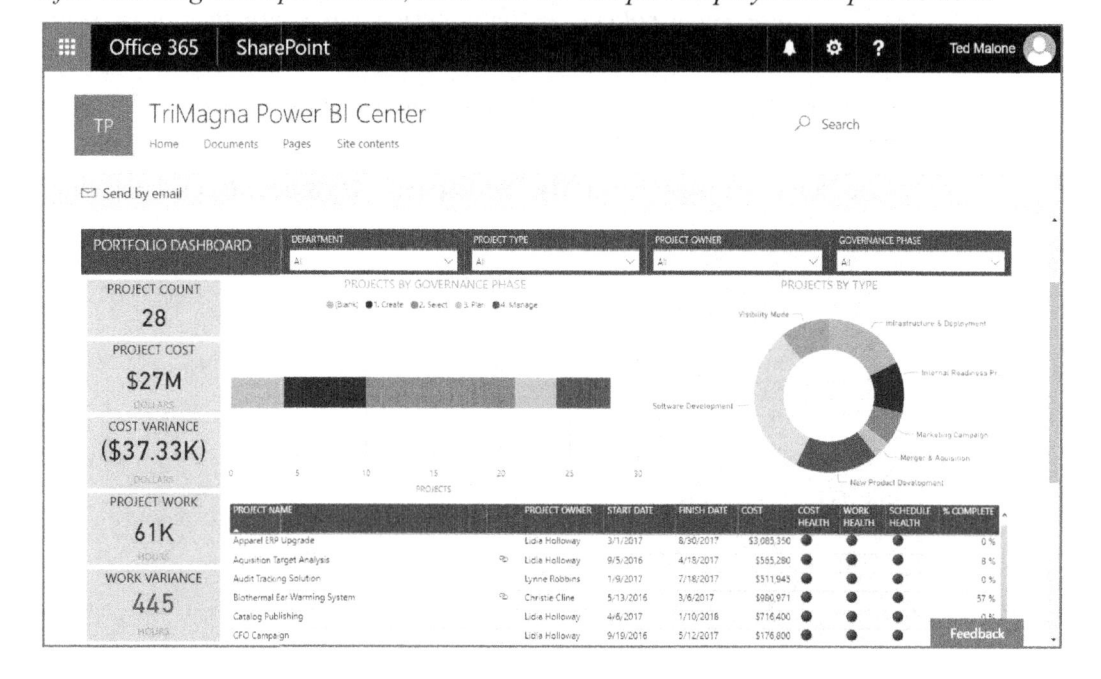

Summary

In this Hands-On Lab, you learned how to create a custom SharePoint Site Collection using Modern UI to host Power BI reports. You also learned how to create a new SharePoint page to be used to host a Power BI report and then added the Power BI web part to the page and embed the link provided from Power BI Pro.

You learned how to sign up for a Power BI (free version) and then upgraded to Power BI Pro and used the Project Online Content Pack to download the built-in dashboards and reports that used the data from the Project Online link to PWA.

You also learned how to modify the Power BI Center home page and shared the site and Power BI reports & dashboards with members of a security group and a user.

Objectives

After completing this Hands-On Lab, you were able to:

- Create a Modern UI SharePoint Site Collection
- Sign Up for a Power BI Account
- Use the Power BI Project Online Content Pack
- Upgrade Free Power BI account to Power BI Pro
- Add Power BI Reports to a SharePoint Page
- Modify the Power BI Center Home Page
- Share the Power BI Center Site
- Share the Power BI Dashboard and Testing

This page is intentionally left blank

Made in the USA
Las Vegas, NV
11 September 2021